Modern

DATE DUE

WITHDRAWN

BRODART, CO.   Cat. No. 23-221-003

Modern Wiring Practice
Design and Installation

Revised edition

W.E. Steward and R.A. Beck

Edited by

T.A. Stubbs

With additional contributions by

W.P. Branson

AMSTERDAM • BOSTON • HEIDELBERG • LONDON
NEW YORK • OXFORD • PARIS • SAN DIEGO
SAN FRANCISCO • SINGAPORE • SYDNEY • TOKYO

Newnes is an imprint of Elsevier

ELSEVIER

Newnes

Newnes is an Imprint of Elsevier
Linacre House, Jordan Hill, Oxford OX2 8DP, UK
The Boulevard, Langford Lane, Kidlington, Oxford OX5 1GB, UK
30 Corporate Drive, Suite 400, Burlington, MA 01803, USA

First edition 1952, Fourteenth edition published 2010

British Library Cataloguing in Publication Data
A catalogue record for this book is available from the British Library

Library of Congress Cataloging-in-Publication Data
A catalog record for this book is availabe from the Library of Congress

ISBN–13: 978-1-8561-7692-7

For information on all Newnes publications visit
our web site at books.elsevier.com

Printed and bound in Italy
10 11 12 13 14 10 9 8 7 6 5 4 3 2 1

Working together to grow
libraries in developing countries

www.elsevier.com | www.bookaid.org | www.sabre.org

ELSEVIER   BOOK AID International   Sabre Foundation

Contents

Part I
Design of Electrical Installation Systems

16. Inspection and Testing

Preface

This book surveys the broad spectrum of electrical design and installation work, and this edition has been revised to incorporate the latest amendments to BS 7671 (The IEE Wiring Regulations) issued in 2008. The book is intended to supplement the various regulations and items of legislation. It is not a replacement for them.

The book is divided into two sections: (1) design of electrical installation systems and (2) practical installation work. The design section, which has been completely revised to reflect current practice, explains in simple terms the various regulations and requirements and goes on to deal with such matters as the fundamental principles, the design process, installation design, distribution and a design worked example.

The practical section, dealing with the most important wiring systems, is based on the authors' experience, and includes many on-site diagrams and photographs. The authors hope that readers will gain much useful information from the book. Any comments on the new edition will be most welcome.

R.A. Beck
T.A. Stubbs

Acknowledgements

We are grateful to many people for assistance with the preparation of this work: firstly, to the Institution of Engineering and Technology for much helpful advice, and for permission to publish extracts from the Wiring Regulations. The Regulations are published as a British Standard, BS 7671, and we are equally indebted to the British Standards Institution for their permission to publish extracts. This book is not a replacement for the IEE Regulations, and copies of these and the guidance notes which accompany them may be obtained from the Institution at Michael Faraday House, Six Hills Way, Stevenage, SG1 2AY.

Many companies and individuals in the field of electrical design and installation work have been instrumental in assisting and giving advice which has helped in the preparation of this edition. We would particularly like to acknowledge contributions from Amtech Power Software, the British Standards Institution, the Chartered Institution of Building Services Engineers, Cooper Lighting and Safety, M W Cripwell Ltd, the Institution of Engineering and Technology, Inviron, W.T. Parker Ltd, Relux Informatik AG., and Wrexham Mineral Cables. Our numerous questions have been answered fully and courteously and this help has enabled me to present a practical and up-to-date volume. Many of the on-site photographs have been possible thanks to the agreement of individual electricians and designers, to whom we are most grateful.

In addition to the above we would like to thank a number of electrical equipment suppliers and individuals who have kindly supplied illustrations and photographs. These are individually credited.

To one and all, we extend our appreciation and thanks.

William Edward Steward was a remarkable man in many ways. Trained as a premium electrical apprentice with Mann Egerton at Norwich, he became an electrician, foreman and, later, the branch manager at the firm's London office. In 1933, acting on advice from his brother, he founded William Steward and Company, engaged on a range of mechanical and electrical contracting works. In the early days he was company secretary, accountant, chief engineer, estimator, electrician, gas fitter and van driver! The firm became a limited company in 1935 and in 1939 was accepted as a member of the Electrical Contractors Association.

By adapting readily to changing trading conditions, the business was kept busy during the war and continued its growth in the years which followed. The company grew steadily from its early days and employee numbers reached 50 in the 1940s, 100 in the 1950s and over 500 by 1975. Many prestigious contracts were undertaken and by 1985, the company had branches in London,

Birmingham, Ipswich, Leeds, Manchester, Norwich, Southampton, Stroud and Walsall, as well as a number overseas.

William Steward died in 1984 and in 1992 the company was sold and became part of the European electrical giant ABB, being renamed ABB Building Technologies. The blend of personal service and professionalism that had been evident from the earliest days was still a feature of the business. A management buy-out of ABB Building Technologies in 2003 created a new company, Inviron, which continues to undertake electrical and mechanical engineering activities, along with facilities management. It is one of the only (and largest) wholly employee-owned service providers of its kind in the UK.

In the 1950s, William Steward was an important employer and figure in electrical contracting, and showed a deep commitment to the well being and future of the industry and the people who worked in it. It was apparent that a handbook for use by electricians, foremen, managers and designers was needed, and the result was the publication of the first edition of 'Modern Wiring Practice' with William Steward himself as author. The book has continued ever since and is now in its 14th edition. It is pleasing to note that the ethos of William Steward is embraced by Inviron which continues to prosper and whose vision 'to become the most respected building services provider in the UK' is a fitting reflection of the philosophy held by William Steward.

Design of Electrical Installation Systems

Chapter 1

Regulations Governing Electrical Installations

Whatever type of electrical equipment is installed, it has to be connected by means of cables and other types of conductors, and controlled by suitable switchgear. This is the work which is undertaken by the installation engineer, and no equipment, however simple or elaborate, can be used with safety unless the installation has been planned, correctly designed and the installation work has been carried out correctly.

1.1 PLANNING OF INSTALLATION WORK

Like fire, electricity is a very good servant, but if not properly controlled and used it can prove to be a very dangerous master. The need for planned methods of wiring and installation work has long been recognised and all kinds of regulations, requirements, recommendations, codes of practice and so on have been issued. Some are mandatory and can be enforced by law, whilst others are recommendations.

This book deals with the work of the electrical designer and installation engineer and an attempt will be made to present, as clearly as possible, a general outline of the basis of good installation work, including design, planning and execution. References will be made to the various rules and regulations, and copies of these must be obtained and studied.

From what has already been said it should be clear to everyone who intends to undertake any electrical installation work that they must be conversant with all of the recognised standards and practices.

If an uninstructed amateur attempts to paint his house, at the very worst he can make an unsightly mess, but if he decides to install a few additional 'points' in his house, his workmanship might become a positive danger to himself and his family.

When planning an installation there are many things which must be taken into account: the correct size of cables, suitable switchgear, current rating of overcurrent devices, the number of outlets which may be connected to a circuit and so on. These and other matters are explained in the various chapters of this book.

FIGURE 1.1 Regulations. It is essential before designing or installing electrical equipment to obtain and study copies of the relevant British Standards, Regulations and other guidance documents. A selection of these is illustrated here.

The regulations governing electrical design and installation work can be divided into two categories: statutory regulations and non-statutory regulations (Fig. 1.1).

Statutory regulations include:

Type of installation/ activity	Regulation	Administered by
Installations in general (with certain exceptions)	Electricity Safety, Quality and Continuity Regulations 2002 and amendments	Secretary of State
All installations in the workplace including factories and offices	Electricity at Work Regulations 1989 and amendments	Health and Safety Executive
Management and design of installations	Construction (Design and management) Regulations 2007	Secretary of State
Installation practice	Work at Height Regulations 2005	Secretary of State
Electrical equipment	The Low Voltage Equipment (Safety) Regulations 1989	Secretary of State

Type of installation/ activity	Regulation	Administered by
Buildings in general with certain exceptions (Separate Regulations apply in Scotland and N Ireland)	Building Regulations 2000 and amendments	Department for Communities and Local Government

Non-statutory regulations include:

Type of installation	Regulation	Published by
Installations in general (with certain exceptions)	Requirements for Electrical Installations. IEE Wiring Regulations Seventeenth Edition BS 7671: 2008	British Standards Institution and the Institution of Engineering and Technology
Installations on construction sites	BS 7375: 1996	British Standards Institution
Conduit systems	BS EN 61386: 2004	British Standards Institution
Trunking and ducting systems	BS EN 50085	British Standards Institution
Accommodation of building services in ducts	BS 8313: 1997	British Standards Institution
Installations in explosive atmospheres	BS EN 60079: 2003	British Standards Institution
Emergency lighting of premises (other than cinemas and similar premises)	BS 5266: 1999	British Standards Institution
Fire detection and alarm systems in buildings	BS 5839: 2002	British Standards Institution
Protection of structures against lightning	BS EN 62305: 2006	British Standards Institution
Industrial plugs, sockets and couplers	BS EN 60309: 1999	British Standards Institution

Continued

Type of installation	Regulation	Published by
Uninterruptible power supplies	BS EN 62040	British Standards Institution
Earthing	BS 7430: 1998	British Standards Institution

1.2 THE ELECTRICITY SAFETY, QUALITY AND CONTINUITY REGULATIONS 2002

The Electricity Safety, Quality and Continuity Regulations 2002 came into effect on 31 January 2003 and were drawn up with the object of securing a proper supply of electrical energy and the safety of the public. An amendment, effective from October 2006, introduced a number of changes. The regulations replace The Electricity Supply Regulations 1988 and subsequent amendments up to and including those issued in 1998.

The Regulations apply to all 'duty holders' concerned with the supply and use of electrical energy and these include generators, distributors, transmitters, meter operators and others supplying electricity to consumers. They also apply to the agents, contractors and subcontractors of any duty holders.

As with the earlier regulations, parts of the 2002 regulations apply to the supply of electricity to consumer's installations (Regulations 23–29 inclusive) and give the electricity distributor powers to require certain standards of installation before giving or maintaining a supply to the consumer. Regulation 25(2) states that 'A distributor shall not give his consent to the making or altering of the connection where he has reasonable grounds for believing that the consumer's installation fails to comply with British Standard Requirements.'

If any installation is not up to the standard, the distributor may issue a notice in writing to the consumer requiring remedial works to be carried out within a reasonable period. The period required must be stated in the notice. If remedial works are not carried out by the end of the period specified, the distributor may disconnect (or refuse to connect) the supply and, in the event of such disconnection must set out the reasons in a further written notice.

A distributor may also disconnect a supply without giving notice, if such disconnection can be justified on the grounds of safety. In this event the distributor must give notice in writing as soon as reasonably practicable, giving reasons and details of remedial measures required. The distributor shall restore the supply when the stipulated remedial measures have been taken.

If there is a dispute between the distributor and consumer over the disconnection or refusal to connect, which cannot be resolved between them, the matter may be referred to the Secretary of State who shall appoint a suitably qualified person to determine the dispute. Following the determination, the distributor shall maintain, connect, restore or may disconnect the supply as appropriate, subject to any conditions specified in the determination.

1.3 IEE WIRING REGULATIONS – BS 7671

The full title is 'Requirements for electrical installations – The IEE Wiring Regulations – Seventeenth Edition. BS 7671: 2008, and is based upon CENELEC (The European Committee for Electrotechnical Standardisation) Harmonisation Documents formed from IEC (International Electrotechnical Commission) standards. The requirements and some of the actual wordings are therefore similar to IEC standards.

The IEE Regulations are divided into the following parts:

Part 1 Scope, object and fundamental principles
Part 2 Definitions
Part 3 Assessment of general characteristics
Part 4 Protection for safety
Part 5 Selection and erection of equipment
Part 6 Inspection and testing
Part 7 Special installations or locations

There are also 15 appendices, and these are:

Appendix 1 *British standards to which reference is made in the Regulations*
Appendix 2 *Statutory regulations and associated memoranda*
Appendix 3 *Time/current characteristics of overcurrent protective devices and residual current devices* (RCDs)
Appendix 4 *Current-carrying capacity and voltage drop for cables and flexible cords. Tables are included for cables with copper or aluminium conductors*
Appendix 5 *Classification of external influences*
Appendix 6 *Model forms for certification and reporting*
Appendix 7 *Harmonised cable core colours*
Appendix 8 *Current-carrying capacity and voltage drop for busbar trunking and powertrack systems*
Appendix 9 *Definitions – multiple source, d.c. and other systems*
Appendix 10 *Protection of conductors in parallel against overcurrent*
Appendix 11 *Effect of harmonic currents on balanced 3-phase systems*
Appendix 12 *Voltage drop in consumers' installations*
Appendix 13 *Methods for measuring the insulation resistance/impedance of floors and walls to Earth or to the protective conductor system*
Appendix 14 *Measurement of earth fault loop impedance: consideration of the increase of the resistance of conductors with increase of temperature*
Appendix 15 *Ring and radial final circuit arrangements, Regulation 433.1*

In addition to the Regulations themselves, the IEE also publish books of Guidance Notes and these include *on-site* and *design* guides.

The guides provide much additional useful information over and above that contained in the 17th edition of the Wiring Regulations themselves.

This present book is based upon the requirements of the 17th edition of the IEE Regulations, and the following comments on each part are offered for the benefit of readers who are not familiar with the layout and presentation.

Part 1 Scope

The scope of the Regulations relates to the design, selection and erection of electrical installations in and about buildings. The Regulations cover the voltage up to and including 1000V a.c. or 1500V d.c. They also cover certain installations exceeding this voltage, for example, discharge lighting and electrode boilers.

The Regulations do not apply to electrical equipment on ships, offshore installations, aircraft, railway traction equipment, motor vehicles (except caravans) or to the aspects of mines and quarries which are specifically covered by Statutory Regulations or other British Standards.

Object

The Regulations are intended to provide for the safety of persons, property and livestock, against dangers and damage which may arise during reasonable use of the installation. The fundamental principles of the Statutory Regulations are considered satisfied if the installation complies with Chapter 13 of the IEE Regulations.

Fundamental Requirements for Safety

The fundamental requirements enumerated in Chapter 13 of the IEE Regulations form the basis on which the remainder of the Regulations is built. This fundamental requirement is also used in the Electricity Safety Regulations and the Electricity Regulations of the Factories Act, but in slightly different words.

Two aspects which are included in the fundamental requirements are worthy of emphasis. Safety does depend upon the provision of a sound, well thought out, electrical design, and also the expertise of good electricians doing a good, sound job. This latter requirement is expressed in IEE Regulation 134.1.1 which states: 'Good workmanship ... and proper materials shall be used ...'. Another item worthy of note (IEE Regulation 132.12) states that the equipment shall be arranged so as to afford sufficient space for installation and accessibility for operation, inspection, testing, maintenance and repair.

Alterations to Installations

This aspect is worthy of special comment, as there are significant implications in the requirements. The subject is covered in IEE Regulations 131.8 and in Section 633. Any alterations to an existing installation must, of course, comply with the IEE Wiring Regulations, and this includes any part of the existing work which becomes part of the alteration. In addition the person making the alteration must ensure that the existing arrangements are capable of feeding the new part safely. This in practice means that the existing installation must be subjected to tests to ascertain its condition. It is not the duty of the installer to correct defects in another part of the system, but it is his duty to advise the person ordering the work. This advice should be in writing. In practice it may be preferable to start the altered wiring from a new distribution board.

Part 2 Definitions

A comprehensive list of definitions used in the IEE Regulations is contained in Part 2 of the Regulations. These definitions will occur constantly and a clear understanding is necessary in order to plan and execute installations. Some of the terms are given below.

Protective conductor: A conductor used for some measures of protection against electric shock and intended for connecting together any of the following parts: exposed-conductive-parts, extraneous-conductive-parts, the main earthing terminal, earth electrode(s), the earthed point of the source, or an artificial neutral.

Circuit protective conductor (*cpc*): A protective conductor connecting exposed-conductive-parts of equipment to the main earth terminal.

Earthing conductor: A protective conductor connecting the main earthing terminal of an installation to an earth electrode or to other means of earthing.

Equipotential bonding: Electrical connection maintaining various exposed-conductive-parts and extraneous-conductive-parts at substantially the same potential.

PEN conductor: A conductor combining the functions of both protective conductor and neutral conductor.

Functional earth: Earthing of a point or points in a system or in an installation or in equipment, for purposes other than electrical safety, such as for proper functioning of electrical equipment.

Live part: A conductor or conductive part intended to be energised in normal use, including a neutral conductor but, by convention, not a PEN conductor.

Barrier: A part providing a defined degree of protection against contact with live parts, from any usual direction of access.

Bunched: Cables are said to be bunched when two or more are contained in a single conduit, duct, ducting, or trunking or, if not enclosed, are not separated from each other by a specified distance.

Overcurrent: A current exceeding the rated value. For conductors the rated value is the current-carrying capacity.

Circuit breaker: A device capable of making, carrying and breaking normal load currents and also making and automatically breaking, under pre-determined conditions, abnormal currents such as short-circuit currents. It is usually required to operate infrequently although some types are suitable for frequent operation.

Residual Current Device (RCD): A mechanical switching device or association of devices intended to cause the opening of the contacts when the residual current attains a given value under specified conditions.

Exposed-conductive-part: Conductive part of equipment which can be touched and which is not normally live, but which can become live when basic insulation fails (e.g. conduit, trunking, metal enclosures etc.).

Extraneous-conductive-part: A conductive part liable to introduce a potential, generally Earth potential, and not forming part of the electrical installation.

Separated Extra-Low Voltage (SELV): An extra-low voltage system which is electrically separated from Earth and from other systems in such a way that a single fault cannot give rise to the risk of electric shock.

Protective Extra-Low Voltage (PELV): An extra-low voltage system which is not electrically separated from Earth, but which otherwise satisfies all the requirements for SELV.

Basic Protection: Protection against electric shock under fault-free conditions. Note that, for low voltage installations, this generally corresponds to protection against direct contact. (*Direct Contact* was defined in earlier editions of the IEE Regulations as 'Contact of persons or livestock with live parts').

Fault Protection: Protection against electric shock under single-fault conditions. Note that, for low voltage installations, this generally corresponds to protection against indirect contact, this being 'Contact of persons or livestock with exposed-conductive-parts which have become live under fault conditions'.

Part 3 Assessment of General Characteristics

Chapters 31, 33–36 and 51 of the Regulations firmly place responsibility upon the designer of the installation to ensure that all relevant circumstances are taken into account at the design stage. These considerations include the following characteristics:

1. Maximum demand
2. Arrangements of live conductors and type of earthing
3. Nature of supply
4. Installation circuit arrangements
5. Compatibility and maintainability

Part 4 Protection for Safety

This section covers:

Protection against electric shock
Protection against thermal effects, e.g. fire and burns and overheating
Protection against overcurrent
Protection against voltage disturbances

These matters are dealt with in detail in Part 4 of the IEE Regulations, in Chapters 41, 42, 43 and 44 respectively.

Part 5 Selection and Erection of Equipment

This section covers:

Common rules, such as compliance with standards
Selection and erection of wiring systems
Protection, Isolation, Switching, Control and Monitoring
Earthing arrangements and protective conductors
Other equipments, such as transformers, rotating machines etc.
Safety services including wiring, escape and fire protection.

Part 6 Inspection and Testing

The requirements for inspection are covered in Chapters 61–63 of the IEE Regulations. They cover Initial verification of the installation by a competent person, periodic inspection and testing and reporting requirements.

Part 7 Special Installations or Locations

Part 7 of the IEE Regulations deals with special types of installation. The Regulations give particular requirements for the installations and locations referred to, and these supplement or modify the requirements contained in other parts of the Regulations.

Installations and locations covered include bath/shower rooms, swimming pools, saunas, construction sites, agricultural and horticultural premises, caravans and motor caravans and caravan parks. There are also regulations on conductive locations, fairgrounds and floor or ceiling heating installations. The full list and requirements can be found by studying Part 7 of the IEE Regulations.

1.4 THE ELECTRICITY AT WORK REGULATIONS 1989

These Regulations came into force on 1 April 1990 and apply to all electrical systems installed in places of work. Amendments have been issued and related

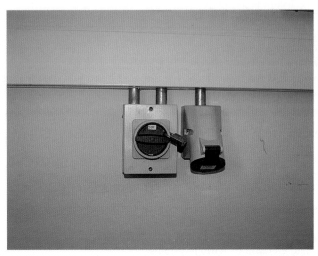

FIGURE 1.2   To comply with the Electricity at Work and IEE Regulations, it is necessary, in appropriate circumstances, to provide means to 'prevent any equipment from being inadvertently or unintentionally energised'. Isolators with provision for padlocking in the isolated position are available to meet this requirement (M.W. Cripwell Ltd).

to explosive atmospheres (1996), offshore installations (1997) and quarries (1999). The Regulations are more wide ranging than the regulations they replace, and they apply to all places of work, including shops, offices etc., as well as factories, workshops, quarries and mines which were covered by previous legislation. They also relate to safety arising from any work activity – not just electrical work – being carried out either directly or indirectly on an electrical system, or near an electrical system.

The Regulations place duties upon all employers, self-employed persons, managers of mines and quarries and upon employees, and cover the construction, maintenance and work activities associated with electricity and electrical equipment. The Regulations come under the jurisdiction of the Health and Safety Commission.

A number of regulations have been revoked or modified as a result of the new legislation and these are listed in full in Schedule 2 of the Electricity at Work Regulations 1989. Some of the main ones are:

The Electricity Regulations 1908
The Electricity (Factories Act) Special Regulations 1944
The Coal and Other Mines (Electricity) Order 1956
The Miscellaneous Mines (Electricity) Order 1956
The Quarries (Electricity) Order 1956

There are 33 regulations in the 1989 edition, and Regulations 4–16 apply to all installations and are general in nature. Regulations 17–28 apply to mines and quarries. Regulations 29–33 cover miscellaneous points. Three books are available from the HMSO which give additional information and guidance and it is recommended that they be obtained and studied. Book 1 covers the Regulations in general, and the other two relate to mines and quarries, respectively.

The Electricity at Work Regulations 1989 imposes a number of new items and there is a change in emphasis in some regulations which significantly alter their application when compared with the regulations they replace. The paragraphs which follow give a brief description of some of the main features.

General No voltage limitations are specified, and the Regulations apply to all systems. Two levels of duty are imposed and these are (1) absolute and (2) as far as is reasonably practicable. The Regulations themselves indicate which level of duty applies to a particular regulation, and further help is given in the *Memorandum of Guidance*.

Regulations 1–3 Introduction These form the introduction, give definitions and state to whom the Regulations apply.

Regulation 4 General This is divided into four parts which cover (1) system design and construction, (2) system maintenance to ensure safety, (3) all work activities on or near the system and (4) provision of protective equipment for persons. All work activities are covered (not just electrical work) and this is sometimes referred to as the 'catch all' regulation. Three of the parts are to be implemented 'as far as is reasonably practicable', but the fourth, on the provision of protective equipment, is absolute. Note that in the definitions a system covers equipment which 'is, or may be' connected to an electrical supply.

Regulation 4(2) refers to system maintenance and it is intended that planned preventative maintenance is used and that the system design is such that this can take place. In this connection it should be noted that adequate working space must be provided. Further details are given under Regulation 15 below.

Regulation 5 Strength and capability Both thermal and mechanical provisions are to be considered, and the arrangement must not give rise to danger *even* under overload conditions. Insulation, for example, must be able to withstand the applied voltage, and also any transient overvoltage which may occur.

Regulation 6 Environments This regulation relates to equipment exposed to hazardous environments, which can be mechanical damage, weather conditions, wet or corrosive atmospheres or from flammable or explosive dusts or gases. There is an important change when compared to the earlier regulations in that the exposure needs to be foreseen, knowing the nature of the activities undertaken at the site, and the environment concerned. This requires a degree of understanding between the designer and the user of the equipment.

Regulation 7 Insulation etc. Requires that conductors be suitably insulated and protected or have other precautions taken to prevent danger. A number of industrial applications will require precautions to be taken to suit the need, where provision of insulation is impractical. For example, with conductor rails of an electrified railway, precautions may include warning notices, barriers or special training for the railway staff. As another example, the use of protective clothing is a requirement of use of electric welding equipment.

Regulation 8 Earthing Requires earthing or other precautions to prevent danger from conductive parts (other than conductors) becoming charged. Metallic casings which could become live under a fault condition are included, and also non-metallic conductors such as electrolyte. Earthing and double insulation are the two most common methods of achieving the requirements, but six others are listed in the *Memorandum of Guidance.*

Regulation 9 Integrity Intended to ensure that a circuit conductor connected to earth or other referenced conductors does not become open circuit or high impedance which could give rise to danger. Reference is made in the guidance notes both to combined and to separate neutral and protective conductive conductors.

Regulation 10 Connections Must be sound, and suitable for purpose, whether in permanent or temporary installations. In particular, connections such as plugs and sockets to portable equipment need to be constructed to the appropriate standards. Also, where any equipment has been disconnected (e.g. for maintenance purposes) a check should be made as to the integrity of the connections before restoring the current, as loose connections may give rise to danger from heating or arcing.

Regulation 11 Excess current protection It is recognised that faults may occur, and protection is needed usually in the form of fuses or circuit breakers to ensure that danger does not arise as a result of the fault. Every part of the system must be protected, but difficulties can arise since in fault conditions, when excess current occurs, it takes a finite time for the protective fuse or circuit breaker to operate. The 'Defence' Regulation 29 applies, and good design, commissioning and maintenance records are essential. The IEE Regulations give further guidance on this subject.

Regulation 12 Isolation Requires provision of suitable means whereby the current can be switched off, and where appropriate, isolated. Isolation is designed to prevent inadvertent reconnection of equipment and a positive air gap is required. Proper labelling of switches is also needed. IEE Regulations 130-06 and 461 are relevant and are described on Page 38 of this book.

Regulation 13 Working dead Precautions to prevent dead equipment from becoming live whilst it is being worked on are required, and can include the locking of isolators, removal of links etc. Isolation, must obviously be from *all* points of supply, so it is a necessity for the operator to be familiar with the system concerned.

Regulation 14 Working live The intention is that no work on live conductors should be undertaken. However, it is recognised that in certain circumstances live working may be required, and the regulation specifies three conditions which must *all* be met before live working is to be considered. Care must be given to planning such an operation, and if live working is unavoidable, precautions must be taken which will prevent injury. It should be noted that the provision of an accompanying person is not insisted upon, and it is for consideration by those involved whether such provision would assist in preventing injury. If accompaniment is provided, the person concerned clearly needs to be competent. In cases where two equal grade persons work together, one of them should be defined as party leader.

Regulation 15 Access Requires that proper access, working space and lighting must be provided. In this connection the contents of Appendix 3 of the *Memorandum of Guidance* should be noted. This refers to legislation on working space and access, and quotes Regulation 17 (of the 1908 Regulations) which should be given proper consideration. In this minimum heights and widths of passageways are specified to ensure that safe access can be obtained to switchboards.

Regulation 16 Competence The object of this regulation is to ensure that persons are not placed at risk due to lack of knowledge or experience by themselves or others. Staff newly appointed may have worked in quite different circumstances, and there is a duty to assess and record the knowledge and experience of individuals.

Regulations 17–28 Mines and quarries These regulations apply to mines or quarries, and separate books of guidance are available from HMSO.

Regulation 29 Defence Applies to specific regulations (which are listed in the Regulations) and provides that it shall be a defence (in criminal proceedings) to prove that all reasonable steps were taken in avoiding the commission of an offence. In applying this regulation it would be essential to maintain proper records and this is relevant for design, commissioning and maintenance matters. Also proper recording of design parameters and assumptions is necessary.

Regulation 30 Exemptions No exemptions have been issued at the time of writing.

Regulations 31–33 General These refer to application outside Great Britain, and to application to ships, hovercraft, aircraft and vehicles. Regulations revoked or modified are also listed.

1.5 BRITISH STANDARDS

Since 1992 the IEE regulations themselves have been issued as a British Standard, BS 7671. In addition, there are many other British Standards which affect electrical installations, and these are designed to encourage good practice. These Standards go into more detail than the other regulations mentioned and IEE Appendix 1 lists those to which reference is made.

1.6 THE LOW VOLTAGE ELECTRICAL EQUIPMENT (SAFETY) REGULATIONS 1989

These regulations impose requirements relating to the safety of electrical equipment. They apply to equipment designed for use at a voltage not less than 50V a.c. and not more than 1000V a.c. (75–1500V d.c.).

The Regulations are statutory and are enforceable by law. They are intended to provide additional safeguards to the consumer against accident and shock when handling electrical appliances. The main requirements are that equipment must be constructed in accordance with good engineering practice, as recognised by member states of the EEC. If no relevant harmonised standard exists, the Regulations state which alternative safety provisions apply.

The requirements state that equipment is to be designed and constructed so as to be safe when connected to an electricity supply and mechanical as well as electrical requirements are specified. If the user needs to be aware of characteristics relevant to the safe use of the equipment, the necessary information should if practicable be given in markings on the equipment, or in a notice accompanying the equipment. Other detailed information is given in the Regulations and in the explanatory notes.

1.7 THE WORK AT HEIGHT REGULATIONS 2005

These regulations impose duties on those carrying out, or responsible for, work at height. In essence, work at height should be avoided whenever practicable. Where access at height is unavoidable, employers must ensure that work activity is planned, supervised and carried out safely. Any situation whereby a fall may result in personal injury is covered and this also extends to the prevention of tools and equipment falling and causing injury to those below (Figs 1.3 and 1.4).

Consideration must be given to the competence of those carrying out the work and other factors such as weather conditions. Risk assessments are required and equipment which is to be used must be properly selected taking into account the access arrangements, frequency of use, tools to be used and the stability of the surroundings. Measures must be taken to ensure that any mobile equipment does not move inadvertently. There are also duties on employees and these relate to safe equipment use, checking and reporting of defects in the equipment.

1.8 HEALTH AND SAFETY AT WORK ACT 1974

The three stages of this Act came into force in April 1975. It partially replaced and supplemented the Factories Act, and the Offices, Shops and Railway Premises Act. It applies to all persons at work, whether employers, employees and self-employed, but excludes domestic servants in private households.

FIGURE 1.3 Battery powered re-chargeable working platforms in use in a new building greatly help in carrying out installation work at heights. The units are stable on level ground, have protection devices fitted and can easily be set by the operator to the most convenient working height (M.W. Cripwell Ltd).

FIGURE 1.4 Another scissors lift suitable for safe working at height. Many industrial and commercial sites require installation work which demands the use of such equipment.

The Act covers a wide range of subjects, but as far as electrical installations are concerned its requirements are mainly covered by those of the Regulations for Electrical Installations, issued by The Institution of Electrical Engineers, and The Electricity at Work Regulations.

The main object of the Act is to create high standards of health and safety, and the responsibility lies both with employers and employees. Those responsible for the design of electrical installations should study the requirements of the Act to ensure that the installation complies with these.

The Health and Safety Executive has issued booklets which give detailed suggestions on various aspects as to how to comply with these requirements. Some of the booklets which mainly affect electrical installations are:

GS 38 Electrical test equipment for use by electricians
HS (G) 38 Lighting at Work
HS (G) 85 Electricity at Work – Safe working practices
HS (G) 107 Maintaining portable and transportable electrical equipment
HS (G) 230 Keeping electrical switchgear safe

The Energy Institute also publishes guidance for petrol filling stations under the title 'Design, construction, modification, maintenance and decommissioning of filling stations'.

1.9 THE CONSTRUCTION (DESIGN MANAGEMENT) REGULATIONS 2007

The Construction (Design Management) Regulations 2007 (CDM 2007) came into force on 6 April 2007 and apply to all construction works within the UK. These Regulations impose a framework of duties on all parties involved in a construction project and it is the responsibility of the designers to familiarise themselves with the requirements of the CDM Regulations and to apply them to the design process. In the context of these Regulations, 'Design' relates to new build, alteration, repair, maintenance, use and decommissioning of sites and therefore it is important that the design activity is comprehensive in all these facets.

There are many roles and definitions involved in the CDM process and these can be found in the text of CDM 2007 itself, in the Approved Codes of Practice and guidance documents. The definition of a designer under CDM 2007 is quite wide and is: 'Any person (including a client, contractor or other person referred to in CDM 2007) who in the course or furtherance of a business either prepares or modifies a design; or arranges for or instructs someone under their control to do so.' Thus the designer is any person (or organisation) that makes a decision that will affect the health and safety of others.

Examples of relevant decisions would be consideration of how much space is allowed to enable the services to be installed and maintained safely; how much time that a line manager allows for the design to be co-ordinated effectively; whether the specification of the materials allowed for by the quantity surveyor in the cost plan is sufficient, and so on.

CDM places some absolute duties on designers and therefore a designer must:

1. Ensure that the clients are aware of their duties,
2. Make sure that the designer is competent for the work undertaken. This includes having adequate resources to enable the design to be completed considering all the heath and safety factors that may be involved,
3. Co-ordinate their work with others to manage and control risks,
4. Co-operate with the CDM co-ordinator (in cases where such a CDM co-ordinator is required) and
5. Provide adequate information about any significant risks associated with the design for the health and safety file.

Following the above, the designer shall avoid foreseeable risks when carrying out design work, for the construction, maintenance and demolition of a structure. In essence this relates to general good health and safety practices, taking reasonable care when designing an installation and using common sense to ensure that no unnecessary risks are taken during construction, maintenance or decommissioning of the electrical systems.

The preference is to firstly eliminate risks 'so far as is reasonably practicable' (SFAIRP) by designing them out. This is covered in CDM Regulation 7. If this cannot be achieved then the next course of action is to reduce the risk to a more practicable level. If any residual risks remain, then reasonable steps must be taken to ensure that they are managed correctly.

One procedure for dealing with potential hazards is use of the acronym 'ERIC' which relates to:

E – Eliminate   R – Reduce   I – Inform   C – Control

As an example, consider an installation involving the provision of lighting at high level, which may introduce hazards from falls etc. when installing/maintaining/removing the fittings. Using the acronym:

E – Eliminate
Consider whether lighting is needed at high level; could it be designed so that the area is lit from a lower level?

R – Reduce
If high level equipment is essential, the potential hazard may be reduced by proposing alternative methods such as the use of remote control gear, or access from a permanent safe working platform.

I – Inform
Relevant information needs to be provided to other designers, the CDM co-ordinator and the person carrying out the work. The information needs to be clear and concise and concentrate on significant risks, some of which may not be obvious.

C – Control

If working at height is inescapable, then it will be necessary to consider a safer means of access than ladders. If the use of ladders is unavoidable, then the design must make appropriate provision for their safe use by providing accessible ladder securing points or allowing for special access equipment such as Mobile Elevated Working Platforms, complying with Schedule 1 of the Work at Height Regulations 2005.

Further guidance on CDM and the use of 'ERIC' is provided at the construction industry training website: www.cskills.org/cdm

Note that a designer is not required to control risk on the site, but must influence factors within his control. They also cannot account for future uses and should not specify the actual construction methods to be used. But in addition to the requirements of CDM 2007, designers must comply with their duties under the Health and Safety at Work Act 1974 and other relevant legislation.

1.10 BUILDING REGULATIONS 2000

These are statutory regulations, and must be complied with, failure to comply with the building regulations may result in an enforcement notice being served. Compliance can be demonstrated in a number of ways. The actual interface with the Building control officer is usually the responsibility of others, but electrical designers and installers have a duty to provide the information they require for their submission for building control approval.

There are multiple parts to the building regulations, and these are:

Part A – Structure
Part B – Fire Safety
Part C – Site Preparation and Resistance to Moisture
Part D – Toxic Substances
Part E – Sound Insulation
Part F – Ventilation
Part G – Hygiene
Part H – Drainage and Waste Disposal
Part J – Combustion Appliances and Fuel Storage
Part K – Protection from Falling, Collision and Impact
Part L – Conservation of Fuel and Power
Part M – Disabled Access to and Use of Buildings
Part N – Glazing
Part P – Electrical Safety

Note that there is an approved person scheme to allow self-certification of building regulation approval for Parts P and L.

The main parts that are of relevance to the design of an electrical installation are Parts L and M, together with Part P which relates not only to the design but also the installation within domestic dwellings.

Part L – Conservation of Fuel and Power

Part L sets targets for maximum carbon dioxide emissions for whole buildings. The regulations apply both to the construction of new buildings and renovation of existing buildings with a total surface area of over 1000 m². For new buildings a net reduction of 40% is often used as an indicator of improvement. Building log books are a legal requirement for new and refurbished non-domestic buildings.

The document is divided into four parts:

L1A: Conservation of fuel and power (New dwellings)
L1B: Conservation of fuel and power (Existing dwellings)
L2A: Conservation of fuel and power (New buildings other than dwellings)
L2B: Conservation of fuel and power (Existing buildings other than dwellings)

Flexibility is permitted as to how the target emissions rates are achieved. This could be by the use of more thermally efficient fabric, more efficient plant or the use of renewable micro-generation. The electrical designer would be mainly concerned with building services, and to achieve the standards required, may be required to make changes to assist the building designer achieve the targets.

Compliance is demonstrated by calculating the annual energy use for a building and comparing it with the energy use of a comparable 'notional' building. The actual calculation is to be carried out either by an approved simulation software, or using a simplified computer program called SBEM – Simplified Building Energy Model which calculates energy use and carbon dioxide emissions from a description of the building geometry and its equipment.

Three parts that affect the electrical installation of Part L2 are:

(i) Controls

This particularly applies to Heating, Ventilation and Air Conditioning (HVAC) controls.

(ii) Energy metering

Most buildings have incoming meters for billing purposes but sub-metering should also be considered as this contributes to good energy management. The strategy for energy metering in a building should be included in the building log book. A reasonable provision would be by installing energy metering that enables 'at least 90% of the estimated energy consumption of each fuel to be assigned to the various end-users. Further guidance is given in CIBSE guide TM39.

(iii) Lighting efficiency

Areas covered are the effective use of daylight, selection of lamp types, lighting control gear, power factor correction, luminaire efficiency and the use of lighting controls. Part L requires that energy efficient lighting be used in both domestic and non-domestic buildings. More advanced solutions include using high frequency dimmable control gear linked to photocells to provide constant illumination with daylight linking. Display Lighting should be switched separately to ensure that it can be turned off when not required.

Part M – Disabled Access to and Use of Buildings

Part M of the Building Regulations 2000 requires reasonable provision to be made to enable people to gain access to and use a building and its facilities. It includes guidance for people with visual and physical disabilities. Part of the section is devoted to the position of switches, outlets and controls. When an electrical installation is being designed and installed, consideration must be given to the ease of identification and use. All users, including those with visual and physical impairments, should be able to locate a control, recognise the settings and be able to use it.

Section 8 of the document deals with accessories, switches and socket outlets in dwellings. The section sets out the heights from floor level of wall-mounted switches, socket outlets and any other equipment in habitable rooms, to enable persons with physical disabilities who have limited reach to be able to operate them. It is usual to demonstrate compliance with this by producing mounting height drawings detailing the access facilities.

Part P – Electrical Safety

Part P of the Building Regulations 2000 relates to electrical installations in dwellings such as houses and flats and their associated areas. The aim of the regulations is to ensure that all modifications and installations to these premises will be carried out by competent persons and in line with the requirements of the IEE and other regulations and guidance (as stated previously within this chapter).

Not all work carried out will fall under these requirements, for example, the replacement of accessories and damaged cables, the installation of lighting points to an existing circuit or main or supplementary equipotential bonding, provided certain conditions are met and they do not involve a special location.

Any work proposed that does fall under the requirements must be notified to the relevant building control body before work begins and such work includes the provision of new circuits, work within special locations (including the replacement of accessories) or a kitchen. The work will then need to be inspected and tested by the local authority.

If the work is carried out by a company or individual that is approved under an approved competent person scheme, then the work need not be notified to Building control, and the company or individual will be able to issue a minor work certificate as a self-certified competent person.

It is the householder who is ultimately responsible for ensuring that any work complies with the building regulations, although the person actually carrying out the work is responsible for ensuring that the works achieve compliance and failure to do so can result in enforcement notices being served and fines for non-compliance, so it is important that the public is made aware of these requirements and any works carried out is in accordance with the regulations.

Other Parts of the Building Regulations

In addition to Parts L and M described above, some other parts may affect the design and installation in less obvious ways. The information which follows is included as these aspects need to be considered by an electrical designer or installer.

Part A – Structural changes to the building which could include the chasing depths of walls, size of penetrations and any other structural changes.

Part B – Fire safety of electrical installations, the provision of Fire Detection and Alarms systems. Fire resistance of penetrations through walls and floors.

This includes the use of thermoplastic materials in luminaire diffusers which form part of the ceiling. Thermoplastic (TP) materials are of two types; Diffusers classified as TPa construction have no restriction on extent of use whereas those classified as TPb construction have limitations on size, area of coverage and spacing. If TPb materials are to be used, careful reference to Part B will be required to ensure that the regulations are met in full.

Part C – Moisture resistance of penetrations.

Part E – The resistance of the passage of sound through floors and walls. Any modifications to the building structure may degrade the resistance to the passage of sound.

Part F – The ventilation rates of dwelling, including use of extract fans.

Chapter 2

Fundamental Principles

2.1 PROTECTION FOR SAFETY FUNDAMENTAL PRINCIPLES

Electrical installations pose a number of inherent risks, which may result in damage or injury to property or its intended users, either in the form of persons or livestock. Regulations set out a number of fundamental principles that are intended to protect against these risks, and it is essential that anyone involved in electrical installation work should understand these principles.

The fundamental principles to be applied to an electrical installation are covered in Chapter 13 of the IEE Wiring Regulations.

IEE Chapter 13 effectively 'sets the scene' and covers the principles to be followed to provide protection for the safety of those that might be affected. It also defines the process to be followed from the design of the installation, through to the selection of the equipment, its installation, verification and testing, to ensure that the requirements of the standard have been met.

More detail is covered in Part 4 of the IEE Regulations. IEE Chapters 41–44 refer to different aspects of the topic and the application of the measures listed in the Regulations.

The areas covered are:

- Protection against electric shock,
- Protection against thermal effects,
- Protection against overcurrent, both overload and short circuit and
- Protection against voltage disturbances and electromagnetic disturbances.

Note: To enable the reader to refer to the relevant parts of the IEE Wiring Regulations more easily, references to relevant parts of the Regulations are enclosed within square brackets as [IEE Regulation 131.2].

Chapter 41 – Protection Against Electric Shock [IEE Regulation 131.2]

The IEE definition of electric shock is 'A dangerous physiological effect resulting from the passage of an electric current through a human body or livestock.' The value of the shock current liable to cause injury depends on the circumstances and individuals concerned. Protection must be afforded in normal service and in the case of a fault. These are referred to in the IEE Regulations as basic protection (formally referred to in the 16th edition as 'protection against "direct contact"') and fault protection (formally referred to as 'protection against "indirect contact"').

A number of the protective measures listed apply to both basic and fault protection, and others apply to one of these only.

Basic protection can be achieved by either preventing the current from passing through a body or by limiting the value of the current to a non-hazardous level.

Fault protection can be achieved by similar methods, but also by reducing the time the body is exposed to the fault, and therefore aiming to reduce the time to a non-hazardous level.

The protective measures available consist of:

- Automatic disconnection of supply (ADS),
- Double or reinforced insulation,
- Electrical separation,
- Extra-low voltage and
- Additional protection.

Automatic Disconnection of Supply (ADS) [IEE Regulation 411]

This protective measure provides both basic and fault protections. Basic insulation, barriers or enclosures to live parts are specified to provide basic protection. Protective earthing, equipotential bonding and ADS under fault conditions provide fault protection. Additional protective measures may also be required by the application of a Residual Current Device (RCD).

Note that in the previous editions of the wiring regulations, the term 'EEBAD' (earthed equipotential bonding and ADS) was used, but this method only related to the *fault* protection, not both fault and basic protections.

Basic protection [IEE Regulation 411.2]: To achieve basic protection the electrical equipment must either employ the requirements of the basic insulation of live parts, barriers or enclosures [IEE Regulation 416] or by the use of obstacles or placing out of reach [IEE Regulation 417].

Protection by insulation [IEE Regulation 416]: is the most usual means of providing basic protection (protection against direct contact) and is employed in most installations. Cables, electrical appliances, and factory-built equipment to recognised standards will normally comply with the requirements but it should be noted that paint or varnish applied to live parts will not provide adequate insulation for this purpose. Basic protection can also be afforded by the use of barriers or enclosures to prevent contact with live parts.

Protection by obstacles or placing out of reach [IEE Regulation 417]: Protection against shock can sometimes be achieved by the provision of obstacles, which prevent unintentional approach or contact with live parts. These may be mesh guards, railings etc. Another method of protection is to place live parts out of arms reach. This is defined in diagrammatic form in Fig. 417 of the IEE Regulations. These two methods may only be applied in industrial-type situations in areas which are controlled or supervised by skilled persons. An example would be the exposed conductors for supply to overhead travelling cranes.

Fault protection [IEE Regulation 411.3]: The object is to provide an area in which dangerous voltages are prevented by bonding all exposed and extraneous conductive parts. In the event of an earth fault occurring outside the installation, a person in the zone concerned is protected by the exposed and extraneous conductive parts in it being electrically bonded together and so having a common potential. The same is not true when a fault occurs within the installation.

The practicalities of the bonding requirements, calculation of the sizes of bonding and protective conductors, and their installation are dealt with in Chapter 4 of this book.

For the protection to be effective it is necessary to ensure that automatic disconnection takes place quickly to limit the duration that a potentially hazardous fault condition could exist. This aspect is covered in IEE Regulation 411.3.2. This is to be provided in accordance with the type of system earthing used, i.e. TN, TT or IT. Tables 41.2–41.4 of the IEE Regulations give the values of earth fault loop impedance for the different conditions and types of protection used to achieve the required Disconnection times.

For TN and IT systems, disconnection times between 0.04 and 0.8s are tabled (depending upon the nominal voltage to earth) for final circuits not exceeding 32A, which includes not only socket outlets but also lighting circuits and others.

The disconnection may be extended to 5s (TN) or 1s (TT) for distribution circuits and other circuits not covered by the final circuits are not covered by the table (IEE Regulation 411.3.2.2 and Table 41.1).

Automatic disconnection is generally brought about by the use of the overload protection device. To achieve a sufficiently rapid disconnection the impedance of the earth loop must be low enough to give the disconnection time required. An alternative way of doing this is by the use of an RCD. The use of an RCD is referred to in the IEE Regulations as Additional Protection.

Additional protection [IEE Regulation 411.3.3]: This is intended to provide protection in the case of the failure of the basic or fault protection and to account for the carelessness of users. Regulation 411.3.3 makes reference to this by introducing additional protection (RCDs) for socket outlets not exceeding 20A, unless for example it is under the supervision of skilled or instructed persons, or it is identified for connection to a particular item of equipment.

Other methods that fall under ADS are the use of Functional Extra-Low Voltage (FELV) [IEE Regulation 411.7] and reduced low voltage systems [IEE Regulation 411.8].

Double or Reinforced Insulation [IEE Regulation 412]

This protective measure is to stop dangerous voltages on accessible parts of electrical equipment if a fault occurs in the basic insulation. **Double insulation** is when additional insulation is implemented to provide fault protection in addition to the insulation which is providing the basic protection, such as double insulated single cabling. **Reinforced insulation** provides basic and fault protection between live and accessible parts, such as insulating enclosure providing at least IPXXB (finger test) or IP2X containing the parts with basic insulation.

Protection by the use of Class 2 equipment: This is equipment having double or reinforced insulation, such as many types of vacuum cleaner, radio or TV sets, electric shavers, power tools and other factory-built equipments made with 'total insulation' as specified in BS EN 60439-1.

Conductive parts inside such equipment shall not be connected to a protective conductor and when supplied through a socket and plug, only a two-core flexible cord is needed. Where two-pin and earth sockets are in use it is important to ensure that no flexible conductor is connected to the earth pin in the plug. It is necessary to ensure that no changes take place, which would reduce the effectiveness of Class 2 insulation, since this would infringe the BS requirements and it could not be guaranteed that the device fully complies with Class 1 standards.

Electrical Separation [IEE Regulation 413]

This protective measure is achieved by providing electrical separation of the circuit through an unearthed source meaning that a fault current to the earth is unable to flow. This means of protection is usually used only where other means of protection cannot be implemented. There are inherent risks associated with this measure, which are increased if supplying more than one item of equipment (in which case IEE Regulation Section 418 must also be met).

Extra-Low Voltage Provided by SELV or PELV [IEE Regulation 414]

These protective measures are generally for use in special locations [IEE Regulations Part 7].

Separated Extra-Low Voltage (SELV) is a means of protection which entails the use of a double wound transformer to BS EN 61558, the secondary winding being isolated from earth, and the voltage not to exceed 50V a.c. or 120V d.c. which can provide both basic and fault protections.

Note that the requirements of the IEE Regulations regarding SELV are modified in respect to equipment installed in bath and shower rooms. The arrangements for provision of switches and socket outlets are relaxed, provided SELV is used at a nominal voltage not exceeding 12V, and provided certain other conditions are met [IEE Regulation Section 701].

In certain circumstances protection may be by extra-low voltage systems with one point of the circuit earthed. This is referred to as Protective Extra-Low Voltage (PELV) but only provides basic protection as the protective conductors of the primary and secondary circuits are connected, and therefore this system may not be used in certain special locations where SELV is allowed.

Additional Protection [IEE Regulation 415]

As mentioned above under ADS, additional protection is usually applied in addition to other protection methods and is required under certain circumstances and in certain special locations. It takes the form of either RCDs and/or supplementary equipotential bonding.

Refer to Chapters 5 and 4 for further information on RCDs and supplementary equipotential bonding, respectively.

Measures Only Applicable for Installations Controlled or Under the Supervision of Skilled or Instructed Persons [IEE Regulation 418]

These measures are also, as the title suggests, only applicable for installations controlled or under the supervision of skilled or instructed persons. They include the provision of non-conducting locations, earth-free local equipotential bonding and electrical separation of the supply for more than one item of current-using equipment. These are special situations and require a number of precautions to be in place before the requirements are met.

Chapter 42 – Protection Against Thermal Effects [IEE Regulation 131.3]

Protection against Thermal effects is covered in Chapter 42 of the regulations, it is especially important as an inherent risk of any electrical installations is its potential to cause fire, either directly or indirectly, therefore it is imperative that measures are employed to ensure that an electrical installation will not present a fire hazard, impair the safe operation of electrical equipment or cause burns to persons or livestock. This chapter is split into three sections:

- Protection against fire caused by electrical equipment,
- Precautions where fire exists and
- Protection against burns.

Protection Against Fire Caused by Electrical Equipment [IEE Regulation 421]

To ensure that any electrical equipment is not liable to cause a fire itself, a number of precautions are required. These include the installation of fixed equipment being carried out in such a way as not to inhibit its intended heat dissipation and in accordance with the manufacturer's recommendations. Luminaires and lamps shall be adequately ventilated, and spaced away from wood or other combustible materials and any potential arcs or sparks that may be emitted in normal services shall be dealt with accordingly.

Fixed equipment containing flammable dielectric liquids exceeding 25L should have provision for safely draining any spilt or surplus liquid, and should be placed in a chamber of fire resisting construction if within a building, adequately ventilated to the external atmosphere.

Precautions Where Fire Exists [IEE Regulation 422]

This section covers the requirements of any electrical services installed in areas where potential fire hazard may exist, due to the materials or processes involved. Interestingly, this also covers the installation of electrical services within escape routes of buildings, the possible spread of fire due to propagating structures (such as cores in high rise buildings) and the enhanced measures to be taken in locations of particular significance, such as museums and national monuments.

Protection Against Burns [IEE Regulation 423]

This section includes a table, which gives the temperature limits for accessible parts of equipment. These range from 55 to 90 °C, dependent on whether the equipment is likely to be touched, and any equipment which will exceed the limits must be guarded so as to prevent accidental contact.

Chapter 43 – Protection Against Overcurrent [IEE Regulation 131.4]

The Electricity at Work Regulation 1989 Part 11 states that 'Efficient means, suitably located, shall be provided for protecting from excess of current every part of a system as may be necessary to prevent danger'. Further, the IEE Regulations state that 'Persons and livestock shall be protected against injury, and property shall be protected against damage, due to excessive temperatures or electromagnetic stresses caused by overcurrents likely to arise in live conductors'. [IEE Regulation 131.4]. These devices can provide either or both overload and fault current protection and could be circuit breakers to BS EN 60947-2: 1996, HRC fuses to BS 88 or BS 1361, or rewirable fuses to BS 3036. Other devices are not excluded from use, provided the characteristics meet one of the afore-mentioned standards.

Overcurrent may be divided into two distinct categories, overload current and fault current.

Overload current is an overcurrent occurring in a circuit which is electrically sound. For example, the current caused by an electric motor which is stalled.

Fault current is that which arises due to a fault in the circuit, as with a conductor which has become disconnected, or in some other way shorted to another, causing a very low resistance fault.

The IEE Regulations deal with overcurrent in Chapter 43 and overload and fault currents are in Sections 433 and 434, respectively. When considering circuit design both aspects of overcurrent have to be taken into account, and it is often possible to use the same device to protect against overload and short circuit. Before doing so it is necessary to determine the design current of the circuit and also to ascertain the prospective short-circuit current which is likely to arise, this is dealt with in IEE Regulations Section 435.

It should also be noted that the neutral conductor shall be protected against short circuits although if the Cross-Sectional Area (CSA) of the neutral conductor is at least equivalent to that of the line conductors, then it is not necessary to provide overcurrent detection and a disconnection device [IEE Regulation 431.2.1] unless it is likely that the current in the neutral conductor may exceed that of the line conductors.

Protection Against Overload [IEE Regulation 433]

The Regulations state that 'every circuit shall be designed so that a small overload of long duration is unlikely to occur' therefore overload protection is intended to prevent the cables and conductors in a circuit from undue temperature rise, and it is necessary to ensure that the rating of the device chosen is suitable for this. Having determined the normal current to be drawn by a circuit, the cable installed must be able to carry at least that value. The protective device in its turn must be able to protect the cable chosen. For example, a circuit may be expected to carry a maximum of 26A. The cable chosen for the circuit must be one which will carry a larger current, say, 36A. The overload device must be rated at a figure between the two so that it will trip to protect the cable but will not operate under normal conditions. In the case quoted a 32A MCB or HRC fuse would be suitable. A device provided for overload protection may be installed at the start of the circuit or alternatively near the device to be protected. The latter is common in the case of electric motors where the overload protection is often incorporated in the motor starter.

In some special circumstances it is permissible to omit overload protection altogether, and [IEE Regulation 433.3] covers this. In some cases an overload warning device may be necessary. An example given is the circuit supplying a crane magnet where sudden opening of the circuit would cause the load on the magnet to be dropped.

In certain circumstances the rated current of the overcurrent protection has to be in effect reduced, for instance circuits supplied by a semi-enclosed fuse (de-rated to 0.725) or directly buried cables (de-rated to 0.9). These aspects are covered in IEE Regulations 433.1.3 and 433.1.4, respectively.

Protection Against Fault Current [IEE Regulation 131.5]

Protection Against Fault Current [IEE Regulation 434]

The Electricity at Work Regulation 1989 Part 5 states that 'No electrical equipment shall be put into use where its strength and capability may be exceeded in such a way as may give rise to danger.' Which is further supplemented by IEE Regulation 131.5 which refers to any conductor being able to carry the fault current without giving rise to excessive temperatures. In addition, any item of electrical equipment intended to carry fault current shall be provided by mechanical protection against electromagnetic stress which could result in injury or damage.

Therefore it is essential that the prospective fault current be known at every relevant point of the installation. Any devices installed shall be capable of carrying the maximum fault current at the point where the device is installed, equally protection shall be provided to interrupt the fault before any conductor or cable permitted limiting temperature is exceeded as this in turn could lead to damage or injury.

There are some exceptions and omissions to these conditions, such as a protective device of a lower breaking capacity than required which is installed downstream of a device of sufficient breaking capacity and is co-ordinated with the device to limit the energy let through to a level which the lower rated device can withstand [IEE Regulation 434.5.1].

Limitation of Currents by the Characteristics of the Supply [IEE Regulation 436]

In a few cases protection is afforded by the characteristics of the supply. Supplies for electric welding come into this category, where the current is limited by the supply arrangements and suitable cables are provided.

Chapter 44 – Protection Against Voltage Disturbances and Measures Against Electromagnetic Influences [IEE Regulation 131.6]

This section of the regulations deals with protection of persons and livestock against any harmful effects as a consequence of faults between live parts of systems at different voltages. The effect of over and under voltages such as lightning strikes, switching or recovery of the circuit from a dip in the supply and providing a level of immunity against any electromagnetic disturbances may influence the installation.

Protection Against Under Voltage [IEE Regulation 445]

Where a danger could arise from a loss in the supply due to reduction in the voltage, an assessment needs to be made of the likelihood of danger arising from a drop in voltage, or loss and subsequent reinstatement of supply, and this may need to be done in conjunction with the user of the installation. Suitable protection may be provided by the use of 'no-volt' or 'low-volt' relays and the IEE Regulations lay down certain conditions for their operation.

Protection Against Power Supply Interruption [IEE Regulation 131.7]

This is a new requirement within the 17th edition of the regulations although designers should have already been considering this, which is basically to ensure that provisions are in place to protect against any danger or damage that could occur if the supply was interrupted.

Additions and Alteration to an Installation [IEE Regulation 131.8]

The last section covered in the Protection for Safety section of the regulations is directed at any alteration or amendments to an installation to ensure that the characteristics of the system are accounted for and that any alteration made will not adversely affect the future operation of the installation, or the rating and condition of any equipment, especially ensuring that the earthing and bonding arrangements are adequate.

2.2 DESIGN FUNDAMENTAL PRINCIPLES

Following on from the fundamental principles for the protection of safety, the IEE Wiring Regulations go on to state, that the electrical installation shall also be designed to provide protection for safety and designed to ensure that the electrical installation shall function correctly for its intended use.

To outline this, the regulations firstly provide the information that is required for the basis of the design, namely:

- The characteristics of the available supply/supplies,
- The nature of the demand,
- Which systems are to be supplied via standby electrical systems or form a system to be used for safety purposes and
- The environmental conditions.

And secondly, outline the requirements that the design shall comply to, namely:

- Considering the conditions the CSAs of conductors shall be determined by:
- The considerations affecting the type of wiring and the method of installation,
- The characteristics affecting the protective equipment,
- The need for emergency control,
- The provision of disconnecting devices,
- The prevention of mutual detrimental influence between electrical and non-electrical installations,

- The requirements of accessibility of the electrical equipment,
- What documentation to be provided,
- The provision of protective devices and switches and
- The provision of Isolation and switching.

Information Required for the Design

Characteristics of Available Supply/Supplies
[IEE Regulation 132.2]

Assessment of General Characteristics [IEE Regulation 301]

Before any detailed planning can be carried out, it is necessary to assess the characteristics of the proposed scheme. This applies whether the installation is a new one, an extension to an existing system or an electrical rewire in an existing building. The assessment includes the purpose and the use of the building, any external influences, the compatibility of the equipment, how it is to be maintained, what systems are required for Safety purposes and what systems need to be continuously operating under a loss of supply. The assessment required is a broad one and some of the aspects to be considered are described below.

Arrangement of Live Conductors and Type of Earthing
[IEE Regulation 312]

A number of aspects of design will depend upon the system of supply in use at the location concerned.

With regard to the arrangement of live conductors, and the type of earthing arrangement, the electricity supplier should normally be consulted for the source of the energy, which in turn can be used to determine the arrangement for each circuit used.

The types of live conductors include single-phase two-wire a.c. and three-phase four-wire a.c. among others.

For the earthing arrangements, five system types are detailed in the IEE Regulations. The initials used indicate the earthing arrangement of the supply (first letter), the earthing arrangement of the installation (second letter), and the arrangement of neutral and protective conductors (third and fourth letters). They are detailed as follows:

TN-S system: A system (Fig. 2.1) having the neutral point of the source of energy directly earthed, the exposed-conductive-parts being connected to that point by protective conductors, there being separate neutral and protective conductors throughout the system. (This is the old system in Great Britain, which is gradually being changed over to a TN-C-S system.)

TN-C-S system: As above (Fig. 2.2) but the neutral and protective conductors are combined in part of the system, usually the supply Protective Multiple

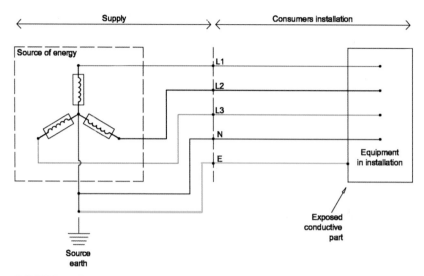

FIGURE 2.1 TN-S system. Separate neutral and protective conductors throughout system. The protective conductor (PE) is the metallic covering of the cable supplying the installations or a separate conductor. All exposed-conductive-parts of an installation are connected to this protective conductor via the main earthing terminal of the installation (E).

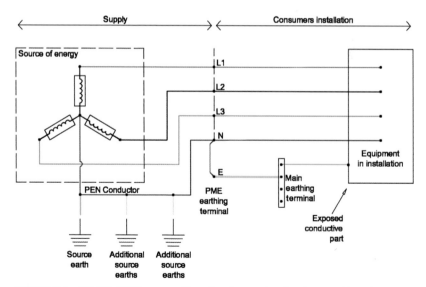

FIGURE 2.2 TN-C-S (PME) system. Neutral and protective functions combined in a single conductor in a part of the system. The usual form of a TN-C-S system is as shown, where the supply is TN-C and the arrangement in the installation is TN-S. The system is also known as PME. The supply system PEN conductor is earthed at two or more points and an earth electrode may be necessary at or near a consumer's installation. All exposed-conductive-parts of an installation are connected to the PEN conductor via the main earthing terminal and the neutral terminal, these terminals being linked together.

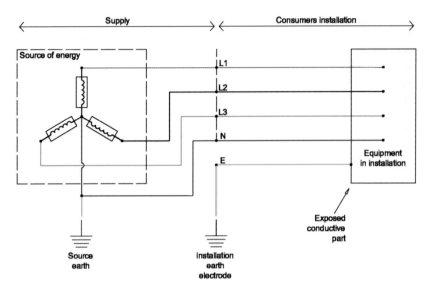

FIGURE 2.3 TT system. All exposed-conductive-parts of an installation are connected to an earth electrode which is electrically independent of the source earth.

Earthing (PME) and then separate in the rest of the system (usually the installation).

TN-C system: As above but the neutral and protective conductors are combined throughout the system and all exposed-conductive-parts of an installation are connected to the PEN conductor. This system is only used in distribution networks as The Electricity Safety, Quality and Continuity Regulations prohibit combination of neutral and protective conductors in a consumer's installation.

TT system (Fig. 2.3): One point of the source of energy directly earthed, but the exposed-conductive-parts of the installation being connected to earth electrodes independent of the earth electrodes of the power system.

IT system: A system where the neutral point of the source of energy is either isolated from earth or connected to earth through a high impedance. The exposed-conductive-parts of the installation are earthed via an earth rod. The electricity companies are not allowed to use this system on the low voltage distribution network to the public.

Three systems are in general use in the United Kingdom at the present time. These are the TN-S, TN-C-S and TT types. The supply undertaking may well provide an earthing terminal at the consumer's installation, and this constitutes part of a TN system. In some cases the protective conductor is combined with the neutral conductor, and in this arrangement the system is the TN-C-S type, the supply being known as PME.

The majority of new systems coming into use are of the TN-C-S type, but before use can be made of the PME type of supply, stringent conditions must be met. Special requirements may apply and the supply undertaking must be consulted. If no earth terminal can be provided by the supply undertaking or the supply does not comply with the conditions for PME, then the TT system must be used.

Another form of TN-C-S is a Protective Neutral Bonding (PNB) arrangement (Fig. 2.4), this generally occurs where the source of supply is provided by the consumer, the District Network Operator (DNO) providing an HV supply only and the transformer being provided by the consumer. The consumer is required to provide the earth, in which case the PNB arrangement allows more flexibility as to where the consumer makes the connection to earth.

Each of the systems described demands different design characteristics and the designer must take such factors into consideration when planning the installation.

The IT system is rarely encountered and will not be found in the United Kingdom as part of the public supply. However, it does have particular application in some continuous process industries, where an involuntary shutdown would cause difficulties with the process concerned. A private supply is necessary and the Regulations require a means of monitoring system faults so that they cannot be left undetected [IEE Regulation 411.6.3].

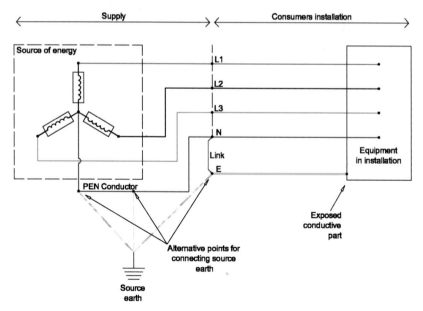

FIGURE 2.4   TN-C-S (PNB) system. Another form of TN-C-S is a Protective Neutral Bonding (PNB) arrangement. This occurs where the District Network Operator (DNO) provides an HV supply only, the transformer being provided by the consumer.

Supplies [IEE Regulation 313]

In most situations it would be necessary to consult with the electrical utilities supplier, such as the DNO to establish the characteristics of the supply, although this can also be achieved by the measurement for an existing supply, or by calculation if for instance the supply is to be taken at High Voltage (HV), and the supply transformer will form part of the customer's installation.

The characteristics to be determined are:

- The nominal voltage and its characteristics including Harmonic Distortion,
- The nature of current and frequency,
- The prospective short-circuit current at the origin of the installation,
- The type and rating of the overcurrent protective device at the origin of the installation,
- Suitability of the supply (including maximum demand) and
- The earth fault loop impedance external to the installation.

If it has been decided to provide a separate feeder or system for safety or standby purposes, the electricity supplier shall be consulted as to the necessary changeover switching arrangements.

FIGURE 2.5 An isolating transformer supplying extra-low voltage socket outlets for use of portable tools in an industrial installation.

Nature of Demand [IEE Regulation 132.3]

It is fairly obvious that when considering the design of an electrical installation, the number and types of circuits to be supplied need to be determined, IEE Regulation 132.3 covers this and states that the information should be determined by the location of the points, the power that is required at those points, what the expected loadings will be, how this is affected by the fluctuations in the demand both daily and yearly and what future demand is to be expected. There also needs to be a consideration for the control and similar requirements as well as any special conditions such as harmonics.

Most of these considerations will need to be addressed when tackling the requirements for the calculation of the maximum demand and diversity.

The Purpose and Intended Use of the Building and the Type of Construction [IEE Regulation 31]

The construction and use of the building will inevitably dictate the type of equipment to be installed, the impact of this is dealt with in Chapter 31 of the IEE Wiring Regulations.

Maximum Demand and Diversity [IEE Regulation 311]

To assess the impact on the existing system and/or the District Network Operators (DNO) available capacity and the viability of proposed system, it is necessary to estimate the maximum current demand. IEE Regulation 311.1 states that diversity may be taken into account. Diversity is not easy to define, but can be described as the likely current demand on a circuit, taking into account the fact that, in the worst possible case, less than the total load on that circuit will be applied at any one time. Additional information is given in the IEE *On-site guide*, Appendix 1. Diversity examples can be found in Chapter 4.

It would also be normal practice when at the planning stage, to make an allowance for future anticipated growth.

Division of Installation [IEE Regulation 314]

Installation circuit arrangements cover the need to divide circuits as necessary to avoid danger, minimise inconvenience and facilitate safe operation, inspection, testing and maintenance. Thus, for example, it is good practice to spread lighting between different circuits so as to prevent a circuit failure causing a complete blackout.

Electrical Supply Systems for Safety Services or Standby Supply Systems [IEE Regulation 132.4]

There are a number of instances where items of equipment will require the supply to be maintained even if the incoming supply has been lost. These generally fall into two categories, those which are required to be maintained for the purpose of the process or end user, such as data installations, critical process lines and the like and those which may be related to life safety systems such as smoke extract, emergency escape lighting, fire fighting lifts or parts of healthcare installations. In either case these need to be identified and the characteristics of the supply (which may be separate to the normal supply) obtained. Examples of safety sources of supply include batteries, generating sets or a separate feeder from the supply network which are independent of each other (such as supplies taken from different primary sub-stations).

When calculating any characteristics associated with these systems, both the normal and stand-by sources need to be considered, and must have adequate capacity and rating plus reliably changeover in the required time [IEE Regulation 313.2].

The specific requirements for stand-by systems are not covered by the IEE Wiring Regulations, but their impact needs to be considered, which is dealt with in Chapter 56 of the IEE Wiring Regulations, further guidance can be obtained from the relevant standards and regulations for the particular system.

Consideration for the continuity of such systems is coved in Chapter 36 of the Regulations.

Environmental Conditions [IEE Regulation 132.5]

Environmental conditions, utilisation and construction of buildings are dealt with in Chapter 51 and Appendix 5 of the IEE Wiring Regulations.

Environmental conditions include ambient temperature, altitude, presence of water, dust, corrosion, flora, fauna, electromagnetic or ionising influences, impact, vibration, solar radiation, lightning and wind hazards. Features in the design will depend upon whether the building is occupied by skilled technicians, children, infirm persons and so on, also whether they are likely to have frequent contact with conductive parts such as earthed metal, metal pipes, enclosures or conductive floors.

Fire risks in the construction of buildings are also covered in IEE Appendix 5. With large commercial premises, these matters may be dealt with by the fire officer, but the electrical designer needs to be aware of such factors as combustible or even explosive contents and also the means of exit.

The Requirements of the Design

As stated previously, as well as the information required referred to above, there are a number of fundamental requirements the design of an electrical installation must meet, the actual design process and how they are incorporated into the design is covered later in this book (Chapter 4), but the principles are outlined below.

CSA of Conductors [IEE Regulation 132.6]

When thinking of the design of an electrical installation, one of the most prevalent issues that comes to mind is the sizing of the cabling supplying a circuit.

The IEE Wiring Regulation stipulates a number of conditions that need to be considered when determining the CSA (or size) of a conductor. These include the admissible maximum temperature, the voltage drop limits, the maximum impedances to operate the protective device and the electromagnetic stresses due to fault currents, how the cables are to be installed and the mechanical stresses the conductor may be exposed to, and what harmonics and/or thermal insulation may be present.

Chapter 52 of the IEE Wiring Regulations covers the majority of the considerations and requirements when determining the size of the conductors including the minimum sizes to be utilised, which is listed Table 52.3 of the Regulations.

Type of Wiring and Method of Installation [IEE Regulation 132.7]

To determine the most suitable type of wiring and method of installation to be utilised, a number of the factors are similar to those considered for determining the CSA of the cabling. Additional considerations include where and how the cabling is to be installed and who or what has access to the wiring as well as likely interference it may be exposed to.

The types of wiring that are currently in use are detailed in Part 2 of this book.

Protective Equipment [IEE Regulation 132.8]

The selection of protective equipment must be in accordance with the fundamental principles that covered earlier in this chapter, namely they be selected to protect against the effects of overload and short circuit, earth faults, over and under/no voltage as applicable, the method of selecting the particular protective device is covered in Chapter 4 of this book, with examples of the types currently in use in Chapter 5 of this book.

Emergency Control [IEE Regulation 132.9]

To ensure that the supply can be interrupted immediately in case of danger arising, emergency control of the supply shall be provided; the requirements of this are covered by IEE Regulation 537.4.2 which states among other requirements that the device must be readily identifiable (preferably red with a contrasting background) and accessible.

Disconnecting Device [IEE Regulation 132.10]

Devices are to be provided to allow switching and/or isolation of the electrical installation and its associated equipment, further application and requirements of these devices are found in Section 537 of the Regulations.

Prevention of Mutual Detrimental Influences [IEE Regulation 132.11]

Mutual detrimental influences include the installation compatibility with other installations (including non-electrical) and Electromagnetic compatibility.

Compatibility [IEE Regulation 331]

The section under the heading of 'Compatibility' deals with characteristics which are likely to impair, or have harmful effects upon other electrical or electronic equipment or upon the supply. These include:

- Possible transient overvoltages,
- Rapidly fluctuating loads,

- Starting currents,
- Harmonic currents (e.g. from fluorescent lighting),
- Mutual inductance,
- d.c. feedback,
- High-frequency oscillations,
- Earth leakage currents and
- Need for additional earth connections.

Suitable isolating arrangements, separation of circuits or other installation features may need to be provided to enable compatibility to be achieved. Rapidly fluctuating loads or heavy starting currents may arise in, for example, the case of lift motors or large refrigeration compressors which start up automatically at frequent intervals, causing momentary voltage drop. It is advisable for these loads to be supplied by separate sub-main cables from the main switchboard, and sometimes from a separate transformer.

Electromagnetic Compatibility [IEE Regulation 332]

Regulations regarding Electromagnetic Compatibility (EMC) came into effect in January 1996. These regulations (The Electromagnetic Compatibility Regulations: 1992) are derived from European Community directives and require that all electrical equipments marketed in the community meet certain standards. The aim of the regulations is to reduce the electromagnetic interference from equipment such that no harmful effects occur; also the equipment itself must be designed to have inherent immunity to any radiation it receives.

Provided equipment satisfies the requirements, the CE mark may be affixed by the manufacturer. Two methods are available to demonstrate compliance. These are the standards-based route, requiring tests by an independent body, or the construction-file route where the manufacturer records the design and the steps taken to achieve compliance. In practical terms the aspects which need to be taken into account include the selection of components, earthing arrangements, equipment layout and design of circuits, filtering and shielding.

Installation engineers will be seeking to use equipment which carries the CE mark but it is important to bear in mind that the installation practice itself has a bearing on the compatibility of equipment. It is possible that the installation arrangements may fail to preserve the inherent EMC characteristics of the individual items of equipment. With large installations or on complex sites, a systematic approach to EMC will be needed. Possible remedies could include the installation of additional screening or filtering, or require cables to be rerouted to address the situation.

A book such as this cannot address this complex subject in detail but specialist advice can be obtained from consultants in this field of work.

Accessibility of Electrical Equipment [IEE Regulation 132.12]
Maintainability [IEE Regulation 314]

Accessibility and maintainability are covered by IEE Regulations 513.1 and 529.3, respectively. Also, consideration must be given to the frequency and to the quality of maintenance that the installation can reasonably be expected to receive. Under the Electricity at Work Regulations 1989, proper maintenance provision must be made including adequate working space, access and lighting. The designer must ensure that periodical inspection, testing and maintenance can be readily carried out and that the effectiveness of the protective measures and the reliability of the equipment are appropriate to the intended life of the installation. It is necessary to ensure as far as possible that all parts of the installation which may require maintenance remain accessible. Architects need to be aware of these requirements and in commercial buildings it is usually the practice to provide special rooms for electrical apparatus.

FIGURE 2.6 IEE Regulation 513 requires that every item of equipment be arranged so as to facilitate its operation, inspection and maintenance. This switchboard is installed in a room provided for the purpose, with access space both in front and behind the equipment. This situation will allow access for proper maintenance (W.T. Parker Ltd).

In carrying out the design process, a number of decisions will be needed at the 'assessment of general characteristics' stage. It is important to record this data so that when required either during the design stage or afterwards, reference can be made to the original assessment process. The person carrying out final testing will require information as to these design decisions so as to assess the design concept employed in the installation.

Maintainability is also at the core of the CDM regulations, covered in the previous chapter, and further guidance as to determining space for the electrical building services can be found from a number of different sources details of which can be found in Chapter 5.

Documentation for Electrical Installations [IEE Regulation 132.13]

For every electrical installation, a certain amount of documentation is required; this ranges from the distribution board circuit chart (IEE Regulation 514.9.1) to the electrical installation certificate, as detailed within IEE Chapter 63, with examples to be found within IEE Appendix 6. The design is not complete until it has been inspected, tested, verified and signed-off.

Protective Devices and Switching [IEE Regulation 132.14] and Isolation and Switching [IEE Regulation 132.15]

Similar to emergency control and disconnecting devices covered before, protective devices and Isolation and switching are covered in Chapter 53 of the Regulations. This states that suitable means of isolation must be provided so that the supply may be cut off from every installation [IEE Regulations 132.15 and 537]. Also, for every electric motor, efficient means of disconnection shall be provided which shall be readily accessible. Chapter 46 of the IEE Regulations deals with the subject in four categories, these being Isolation, Switching for Mechanical Maintenance, Emergency Switching and Functional Switching. In many cases one device will be able to satisfy more than one of the requirements. Switching for operating convenience, sometimes termed 'functional switching' or 'control switching' is covered in IEE Regulation 537.4 and in certain cases it may be possible to use functional switches as isolating devices, provided they comply with the Regulations.

A distinction exists between Isolation and Switching for Mechanical Maintenance. The former is intended for operation by skilled persons who require the circuit isolated so as to perform work on parts which would otherwise be live, whereas the latter is for use by persons who require the equipment disconnected for other reasons which do not involve electrical work. In both cases it may be necessary to provide lockable switches or some other means of ensuring the circuit is not inadvertently re-energised, but there are some differences in the types of switches which may be used.

Emergency switching is required where hazards such as rotating machinery, escalators or conveyors are in use. Suitable marking of the emergency switch is

TABLE 2.1 An Extract from IEE Table 53.2 Showing Some Switching and Other Devices Permissible for the Purposes Shown. IEE Regulations Section 537 gives additional information on this topic

Device	Use as isolation	Emergency switching	Functional switching
RCD	Yes[a]	Yes	Yes
Isolating switch	Yes	Yes	Yes
Semiconductors	No	No	Yes
Plug and socket	Yes	No	Yes[b]
Fuse link	Yes	No	No
Circuit breaker	Yes[a]	Yes	Yes
Cooker control switch	Yes	Yes	Yes

[a]Provided device is suitable and marked with symbol per BS EN 60617.
[b]Only if for 32A or less.

required, and it is often the practice to provide stop buttons in suitable positions which control a contactor.

Every installation must be provided with means of isolation but the requirement for the other two categories is dependent upon the nature of the equipment in use. In all but small installations it will be necessary to provide more than one isolator so that the inconvenience in shutting down the whole installation is avoided when work is required on one part of it. If insufficient isolators are provided or if they are inconveniently placed then there may be a temptation to work on equipment whilst it is still live (Table 2.1).

Chapter 3

The Design Process

Any wiring installation requires a good deal of forethought before it can be successfully installed. The process of design is a pre-requisite and this activity requires the consideration of a wide range of issues affecting the site and buildings. It is the intention of this text to provide an overview of the systems and process involved for an electrical engineer delivering the design of an installation. Therefore the first port of call is to review the general design process.

The costs associated with an installation and the programme and time implications are of the utmost importance and can prove to be a determining factor. It is not the intention of this text to cover these in any great detail although decisions affected by whole life costing of an installation will play a part. Some of these influences are shown in Fig. 3.1.

3.1 THE STAGES OF DESIGN

When it comes to designing the electrical services there are many approaches that can be taken, each of which has slightly different outcomes. One approach, defined by the Royal Institute of British Architects (RIBA), is in common use. This is known as the 'RIBA Plan of Work' and covers each of the disciplines and the interested parties. In this plan there are 11 sections which cover the three main areas of pre-design, design and construction.

Responsibility for the design in a particular project is usually laid out and agreed at the start of a contract so that individual activities can be defined and each member of the team can concentrate on the areas that fall under his/her responsibility. The work to be completed by the designer will depend upon their role in the project (i.e. client/consultant/contractor/installer) and the type of project being undertaken (i.e. design and build, new build, refurbishment, extension etc.).

In the section which follows, the design approach is divided into five stages as shown in Fig. 3.2, these being broadly based on the RIBA scheme. This is intended to provide a useful guide, outlining the stages of producing an electrical building services design. Not all these stages are relevant to every project and in some cases would be the responsibility of others, such as the consulting engineer or client.

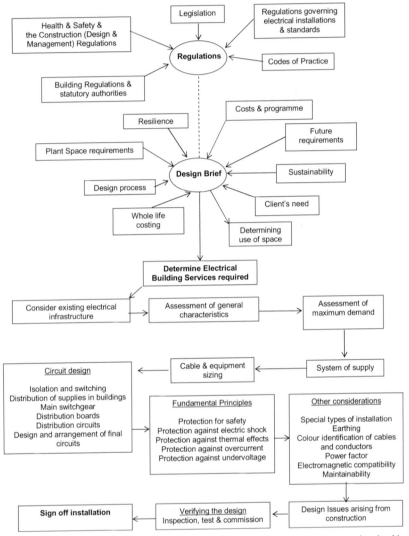

FIGURE 3.1   A typical design map showing some of the considerations and processes involved in the design of the electrical building services (R.A. Beck).

Stage 1 – Preliminary Design Stage

This initial stage is often carried out by the developer, client or architect and may evolve in conjunction with the development of the building outline or by working around a pre-determined plan.

It lays out the basic components of the system, such as defining the electrical system philosophy, the proposed design criteria and extent of the electrical

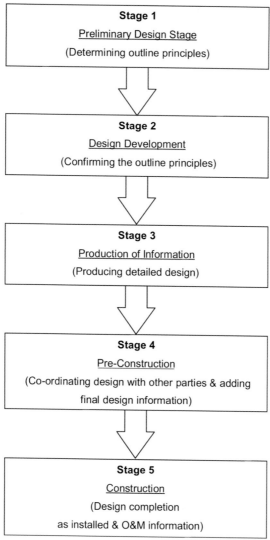

FIGURE 3.2 Stages of the design process from conception to completion (R.A. Beck).

system. The work will be done in consultation with the clients and possibly by their designer, if appropriate for the project. The outcome of this stage is to determine preliminary locations of plant and risers, the concept sketches and schematics to enable the client to judge the feasibility, set high-level principles and to start to integrate the services into the building fabric. This information, along with the preliminary performance specification, will be used to develop the next stage, and is usually carried out by the clients or their appointed consultants.

Stage 2 – Design Development

This stage develops the concepts further into a design, and at this point the aim is to ascertain the spatial requirements of the main cable distribution routes and locations of switchgear. This is achieved by determining the approximate loads for each area and assessing potential equipment requirements and their loadings. System requirements will also need to be considered, so that sufficient allowances can be made for supply change-over, fire detection and the provision of power factor correction equipment. A certain amount of calculation may be required and this will include approximate main cable sizes for sub-main distribution, and an assessment of the anticipated maximum demand of the installation. This will allow preliminary consultation with the District Network Operator (DNO).

The outcome will lead to the production of an outline scope of works and updated concept drawings. The designer's risk assessments should also be underway at this stage. The information produced will set reasonable allowances for the equipment spatial requirements, so that the client may assess the financial and programming aspects of the work.

Stage 3 – Production of Information

Stages 3 and 4 may be carried out by either the consultant or the contractor depending on the terms of the agreement.

Stage 3 is mainly concerned with the production of more detailed information, and the building General Arrangement (GA) drawings which should now have been frozen and include sufficient allowances for the building services, so that the design of the systems may proceed.

The project is now at the stage of carrying out detailed design calculations, including the sizing of cables and containment, the production of a detailed specification and materials and workmanship specification for the electrical services. Detailed drawings, schematics and builders work information can be produced and the building log book can be prepared and preliminary information added. Equipment schedules detailing the actual electrical equipment to be utilised can be provided, including ancillary systems, luminaries and distribution equipment.

Depending on how the project is procured, and how the contractor is appointed to carry out the works, the information produced at this stage may not be fully passed on. In some situations it will serve as a check against the final installation design by the contractor.

Stage 4 – Pre-Construction

Stage 4 is the final stage before construction, and can be carried out either by the contractor installing the works (design and build) or via the more traditional contract where the consultant provides the detailed design information. It is intended at this stage that the previous information is expanded to include more details, and input is gained from specialist designers of systems such as fire detection and alarm system which is incorporated into the design.

Detailed working drawings and spatial co-ordination are checked to ensure that all clashes are minimised, and appropriate corrections are undertaken. Access to plant rooms is detailed and a check is made to ensure that cable entries are acceptable and clearance for maintenance is provided. Information previously provided is verified, including cable sizing, allowable voltage drops, and fault level analysis, evaluating any alternative equipment or plant proposals.

All of this leads to the production of detailed co-ordinated electrical drawings, plans, installation drawings, wiring diagrams, schematics, interface detail schedules, dimensioned drawings of switchgear, commissioning arrangements and circuit protective device settings.

Co-ordinated drawings may be provided and these should include, for example, a reflected ceiling plan detailing the location of all the lighting, fire detection, security, and public address equipment. These, in turn, need to be co-ordinated with the drawings for air conditioning units, grills, fans and other mechanical equipments.

Stage 5 – Construction

The final stage is the actual construction, which includes ongoing design issues that arise from the construction process, verifying the design in practice, and the testing and commissioning of the installation.

Document deliverables from this include producing the testing and commissioning data, log book, as fitted or record drawings, liaison with statutory authorities for the sign off of systems (such as the building control and fire officer) and calculation and data for Operation and Maintenance manuals.

Through all the stages the requirements of the Construction (Design and Management) Regulations 2007 (CDM 2007) will need to be considered.

3.2 THE COMPONENTS OF THE DESIGN PROCESS

Design Brief

Once the design philosophy is known and responsibilities are clear, the detailed design work may proceed. As explained in the foregoing section, the design responsibility may well fall to a single individual on smaller projects, or be divided among several parties on larger works.

The initial stage in the process is to develop the design brief by defining the electrical system philosophy. When carrying out this task, it is essential that a number of items are covered, these include:

- The Client's needs
- Determination of the electrical building services required by the client

- Use of space
- Environmental and installation conditions
- Plant space allowance and spatial co-ordination
- Design Margins

Overriding factors include:

- Regulations and Legislation
- Health and Safety and CDM

Once the brief has been agreed then the electrical building service requirements can be determined so that the design can be developed and the outline principles confirmed, therefore looking at each item in turn:

Client's Needs

It hardly needs emphasising that the Client as the customer is the main influence of the project. No matter whether the project is large or small, the client's requirements need to be satisfied.

These may range from simple to highly complex, for example, they may be as simple as a need for a new hospital. But the achievement of these may demand quite a complex process. The knowledge and experience of the client will also determine the complexity of the project as some clients may be very specific in their requirements, making the process simpler, whilst others may not be skilled in the knowledge of electrical installations. This can add complications in determining the requirements.

Either way the designer and installer have a duty of care to ensure that not only a safe and correctly engineered solution is provided, but also that it meets the need of the client in terms of their operational strategy and the installations functionality and performance.

It should be noted that the IEE Wiring Regulations (BS 7671) state that the Regulations themselves may need to be supplemented by the requirements of the persons ordering the work. An example would be where a particular location may not require additional bonding to meet the requirements of the Regulations and other applicable documents, but it may be required by the actual client process. It is, therefore, important to understand the nature of the client's requirements and any additional considerations that need to be taken into account.

Determination of the Electrical Building Services Required by the Client

At its simplest, the clients (or their representative) could specify their requirements for the electrical building services, such as lighting, small power, fire alarm, public address supplies etc. in a number of forms.

One format could be the preparation of a brief scope of works stating that the electrical installation is to be installed to all relevant British Standards and codes of practice, that the lighting is to be designed to CIBSE standards and an outline requirement for small power, for example, a requirement for a double socket outlet per 10 m^2 of floor area. An alternative would be the preparation of outline sketches of the equipment requirements, the production of room data sheets, a notated drawing or combination of any of the above.

In other cases, the designer will need to consider whether there are any specific client/user requirements such as the supplies to vending machines, kitchen equipment and Mechanical Handling Equipment (MHE). There may be specific supply arrangements, such as the need for stand-by supplies in the form of diesel generators and/or Uninterruptible Power Supply (UPS) systems. There may be other requirements that the client may not request directly, but will require electrical services such as the other building services, including supplies to HVAC plant.

It would then be the responsibility of the designer to interpret these requirements to assess the electrical building services required and then to seek approval from the client that these interpretations are correct. The requirements tend to be based on certain standards, for example, those published by CIBSE for the performance of the lighting, others are based on experience and knowledge from past projects, some may be due to an assessment of risk, such as the need for standby generation and some are more specific requirements by the client.

Use of Space

At this stage, consideration will need to be focussed on the purpose and intended use of the building and the type of construction. What space is to be utilised for? General areas can include offices, ancillary, control rooms, internal distribution and storage, plant areas, WCs or a canteen. Specific areas in a project can include lecture halls, workshops or classrooms in a school.

The construction and use of the building will inevitably dictate the type of equipment to be installed. The characteristics of each area will needed to be considered, such as floor area required, ceiling heights and the floor/wall/ceiling type and construction.

All of the factors above have an impact on the electrical system, if more lighting is required to compensate for the reduced performance of the luminaries due to a poor maintenance factor, then more luminaries may be required. This may lead to more wiring, more circuits and more load, which means increases to the distribution system upstream, which may affect the protective arrangements, it could require more switchgears due to the accumulative effects on the distribution system, so more plant space is required.

The control methods may also need to be revised, for example, the routing of the cabling for the switching may need to be altered if the operation of the building determines that the switch locations are different from those originally assumed.

Installation Conditions

Environmental conditions, utilisation and construction of buildings are dealt with in Appendix 5 of the IEE Wiring Regulations.

Environmental conditions include ambient temperature, altitude, presence of water, dust, corrosion, flora, fauna, electromagnetic or ionising influences, impact, vibration, solar radiation, lightning and wind hazards. Features in the design will depend upon whether the building is occupied by skilled technicians, children, infirm persons and so on, also whether they are likely to have frequent contact with conductive parts such as earthed metal, metal pipes, enclosures or conductive floors. The latest edition of the wiring regulations puts a much greater emphasis on the term skilled/instructed persons, which in turn dictates a number of the new requirements such as the use of enhanced protection where an installation isn't under their supervision.

Fire risks in the construction of buildings are also covered in IEE Appendix 5. With large commercial premises, these matters may be dealt with by the fire officer, but the electrical designer needs to be aware of such factors as combustible or even explosive contents and also the means of exit.

Therefore, as well as the environmental conditions, the expected occupancy, the location and orientation of the installation as well as the external design conditions all require to be considered.

Plant Space Allowance and Spatial Co-ordination

Space requirements for plant access, operation and maintenance of electrical installations should be in accordance with the statuary regulations, codes of practice and guidance documents, in addition to taking into consideration the requirements of CDM. Relevant documents include:

- Management of Health and Safety Regulations
- Workplace (Health, Safety and Welfare) Regulations
- Provision and Use of Work Equipment Regulations
- Construction (Design and Management) Regulations
- Manual Handling Operations Regulations
- Electricity at Work Regulations

Other documents provide guidance or good practices and include those published by the Building Services Research and Information Association (BSRIA). Consideration also needs to be made to the anthropometric data detailed within BS 8313, which gives allowances for the minimum space required to perform certain tasks.

The requirements of the Construction (Design and Management) Regulations 2007 (CDM 2007) have been covered in Chapter 1. Requirements include the need to allow adequate space for installation, access, maintenance and plant replacement, and a means of escape. Fire engineering is required to maintain fire integrity by appropriate compartmentation, access required by local authority and the DNO and the provision of sufficient working space. Heat dissipation from equipment is to be considered, as well as co-ordination with structural elements, restrictions and avoidance of clashes. Some plant may, with advantage, be placed 'on grid lines' to assist with structural loadings.

Maintenance access is an obvious requirement but consideration is also needed as to equipment selection. This can reduce the frequency and the requirements for maintenance and can assist in reducing the overall workload and therefore the amount of time a risk is present. A simple system may be preferable, as unnecessary complication may lead to additional maintenance or specialist care being needed, thus increasing whole life costs. The use of standard equipment and that supplied by reputable manufacturers can ensure future support and simplify the maintenance requirement using common equipment in line with other used on site, reducing the number of different types of equipments in use.

A matter which demands careful consideration is the space requirement and positioning of switchrooms and cable distribution routes. Consideration will need to include bending radius of cable and cross-overs, segregation of services, risers and their sizes, primary distribution routes, plant areas and primary equipment locations. Other matters such as weight loadings, plant support, access, understanding the building as regards to levels, the ceiling and floor void spaces also need attention.

Overall consideration is required for the co-ordination of services. This can be achieved by examining dimensioned drawings of plant space requirements, leading to co-ordinated detailed plant room layouts, encompassing all services and allowing for future requirements and access. Consideration must also be made with respect to the acoustic performance of equipment. For example, allowance must be made for attenuation of generators supply and exhaust air systems.

Design Margins

In carrying out the design process, a number of decisions will be needed at an early (the 'assessment of general characteristics') stage. It is important to record this data so that when required either during the design stage or afterwards, reference can be made to the original assessment process. The person carrying out final testing will require information as to these design decisions so as to assess the design concept employed in the installation.

It is of utmost importance that the design margins are agreed at the start, to allow equipment to be selected correctly and the allowance of plant space to be incorporated into the design. The design must also include allowances for the agreed future expansion. These requirements may be specified within the brief for the project, or decided by the designer based on available design data or previous experience. The experienced designer will be able to make accurate assumptions and agree them so as to minimise the amount of reselection or modification required further into the project. These margins also allow for a certain amount of flexibility later in the design and possible saving or more efficient design decisions.

Allowing future flexibility of the design is more than just allowing excess capacity within the system. It is also allowing flexibility in the utilisation of other equipments or systems, and the possibility of an initial increase capital cost, allowing for a reduced running and maintenance cost and a probable benefit in terms of whole life costing. In addition, future flexibility can add value to the design by facilitating future changes in the use of a building.

Chapter 4

Installation Design

Those responsible for the design of electrical installations, of whatever size, must obtain and study very carefully the requirements of the Wiring Regulations for Electrical Installations, and also statutory regulations, details of which are given in Chapter 1.

The 17th edition of the Wiring Regulations deals with the fundamental principles (covered in Chapter 2 of this book) and gives the electrical designer a degree of freedom in the practical detailed arrangements to be adopted in any particular installation. It is necessary to be sure that the detailed design does in fact comply with the requirements laid down, and as a result a high level of responsibility has to be carried by those concerned with installation planning and design. In many cases, the experience and knowledge of the designer will be called into play to arrive at the best or most economical arrangement and this will encompass the practical application of installation techniques, as well as the ability to apply the theoretical aspects of the work. It will generally be necessary to demonstrate compliance with the Regulations and, in view of this, records need to be kept indicating the characteristics of the installation, the main design calculations and the assumptions made in finalising the design.

4.1 LOAD ASSESSMENT AND MAXIMUM DEMAND

A load assessment is carried out to determine the maximum connected load; this can be achieved by a number of different methods, depending on what needs to be accomplished. Examples are given below:

1 – Estimating the demands by building types and areas on a W/m^2 basis. This method could fulfil the estimation required for Stage 2 of the design process to allow preliminary consultation with the DNO. It is based on rule of thumb data which can be taken from a number of sources such as experience from past projects, specific information provided by the client or published estimates such as CIBSE guide K Table 4.1, an extract of which is shown in Table 4.1 or the BSRIA rule of thumb guidelines for building services, which suggest 10–12W/m^2 for lighting and 15W/m^2 for small power per square metre of office area for instance. Other specialist, mechanical and process loads, will need to be added to these figures, and this method of estimation is useful for checking the site demand figures.

TABLE 4.1 Minimum Design Load Capacities for Lighting and Small Power Equipment for Various Types of Buildings (Reproduced from CIBSE Guide K: *Electricity in buildings*, by Permission of the Chartered Institution of Building Services Engineers)

Building type	Minimum load capacity (W/m^2)
Office	60
School	30
Residential building	30
Hospital	25

2 – Assessing individual systems.   The electrical load can be broken down into a number of categories, such as lighting, small power, Heating Ventilation & Air Conditioning (HVAC), Mechanical Handling Equipment (MHE) and other specialist equipment for instance.

Again there are a number of published sources for this information (such as CIBSE Guide F: Energy Efficiency in Buildings). At the early stages of design, W/m^2 estimates may be used, with outline figures from specialist and other service contractors. Ideally as the design progresses, the actual loading should be utilised to gain a more accurate figure. By extracting quantities from the lighting scheme and referring to manufacturer's details for the luminaires, circuit loadings may be derived and, applying a factor for diversity, the final circuit cable loading can be calculated. An example is shown in Table 4.2.

3 – Imposed load on the distribution system.   Once the outline distribution system is known, the supplies can be broken down into individual distribution areas, so that the actual loading on each main LV panel, panel board, distribution board and sub-main cable can be assessed. This is useful as a check that the correct rating of distribution equipment has been selected.

The above information not only confirms the loading requirements, but can be utilised in determining the metering strategies, load shedding scheme for standby generation (if required) and heat gains on mechanical equipment. In addition, it can be utilised for inputting the design loads for the cable calculations.

Diversity

The initial load schedule is based on a number of sources. Some of these sources may have a certain amount of diversity included, such as the W/m^2 figures, but information gained from manufacturer's data and motor loads will be the actual connected load on the circuit. In this case, the data is based on the maximum electrical load that the particular item may require, meaning the

TABLE 4.2 Typical Electrical Data for Metal Halide Discharge Luminaires (Cooper Lighting)

Nominal lamp (W)	Total circuit (W)	Start current (A)	Mains running (VA)	Power factor
75	81	0.98	89	0.91
100	115	1.00	126	0.91
150	170	1.80	187	0.91
250	276	3.00	325	0.85
400	431	3.50	501	0.86

maximum connected load that is the simple sum of all the electrical loads that are (or may be connected in the case of future provision) connected to the installation.

Diversity occurs because the actual operating load of a system very rarely equals the sum of all the connected loads. This is due to a number of factors such as:

- The general operating conditions of the system. This covers seasonal demands on HVAC systems, or conditions specific to the individual project.
- Time. The fact that not all equipment is operating simultaneously.
- Distribution. The equipment may be spread out over a larger distribution network, whereby operation of equipment at any given point of time is not simultaneous at all locations. In addition loads peak differently and therefore, as the load imposed on the system is the sum of the downstream components, the overall load profile will be lower than the sum of estimated loading of each individual item.
- Actual running conditions. Individual items of equipment may not always run at the rating given on the nameplate of the equipment due to a number of factors, such as variation in load, controls or loadings only required for short periods to meet the operating criteria.
- Equipment not connected. For example, not all the socket outlets on a circuit may have equipment connected to them.
- Intermittent use, such as hand driers, lifts, photocopiers or ancillary areas, only used occasionally.
- Emergency use, such as sprinkler pump sets and smoke extract systems.

A safe design diversity would be 100% of the connected load but in most cases this would not be appropriate and may not be economical. There are a number of factors/conventions that must also be considered, such as the demand factor, load factor or the diversity factor. These are generally all encompassed within the term diversity.

Worked Example

A 6-way 100A single phase & neutral Distribution Board (DB) may have four 32A circuits each supplying a ring circuit of eight 13A sockets. Allowing for the maximum current available at each socket, it would have to have four 104A circuits that would require a 416A supply to the distribution board.

A more reasonable assumption would be to allow, say 2A for each socket outlet, therefore requiring a 16A load per final circuit on a 32A supply (50% diversity factor).

This would in turn impose a diversified load of 64A on the 100A dist. board (a 64% diversity factor). This 'spare' capacity on the circuits and dist. board also acts as a buffer, so that any of the final circuits could have a potential loading of 32A when required, as long as the overall load imposed on the DB wasn't above 100A.

For the maximum demand calculation the 2A per socket could be applied, but for the actual cable calculation, an 80% diversity factor may be imposed on the DB, giving a design load of 80A. The total available capacity of the board (i.e. 4 × 32A) isn't available simultaneously, but there is the capacity to supply the maximum available current of 100A over the four circuits. The figure of 80A will allow for future loadings and fluctuations in the final circuits, while still maintaining a more economical figure to be used in the upstream distribution design.

To produce an economical design of an electrical installation, it will almost always mean that diversity has to be allowed. However, knowing what diversity to allow, and what factors need to be considered is one of the more difficult aspects of the design process. Some guidance is available from the previously mentioned publications such as CIBSE and BISRIA, and further guidance can also be found in the IET Guidance Notes 1 (Appendix H) and other texts. An adequate margin for safety should be allowed and consideration should also be given to the possible future growth of the maximum demand of the installation, which again based on the BISRIA rules of thumb, would be 20%.

To assist with the loadings imposed on a system, control systems or operating procedures may need to be introduced to ensure that the maximum capacity of the systems isn't exceeded, causing penalties from the supply authorities or more severely, a loss of supply due to the operation of the protective device due to overload. For example, such items that may require additional control could be compressed air systems that utilise multiple compressors with large electric motors. These may have high inertia starting loads or inrush currents and if started as a group could cause significant problems to the distribution system. A suitable control system would alleviate these problems by ensuring that the loads do not occur simultaneously.

4.2 CIRCUIT DESIGN

Having outlined some of the main requirements of the IEE Regulations in Chapter 2, it is now proposed to go into the more practical aspects of installation design. The main considerations are to determine the correct capacity of switchgear, protective devices and cable sizes for all circuits. In order to do this it is necessary to take into account the following:

- Subdivision and number of circuits
- Designed circuit current
- Nominal current of protective device
- Size of live conductors
- Determine rating factors:
  - Overload protection
  - Grouping factor
  - Ambient temperature
  - Thermal insulation
- Voltage drop
- Short-circuit protection
- Earth fault protection
- Protection against indirect contact.

Subdivision and Number of Circuits [IEE Regulation 314]

Even the smallest installation will need to be divided into a number of circuits. This is necessary for a number of reasons, including the need to divide the load so that it can be conveniently and safely handled by the cabling and switchgear. To satisfy IEE Regulation 314.01 the designer needs to take into account the likely inconvenience of losing a supply under a fault, facilitating safe inspection, testing and maintenance and reducing unwanted tripping of Residual Current Devices (RCDs), among others.

In small installations it is appropriate to provide a minimum of two lighting circuits, so that in the event of a protective device tripping under fault conditions, a total blackout is avoided. In addition separate circuits may be provided for lighting and power.

In the past there was a tendency to provide insufficient socket outlets, with the result that a proliferation of adaptors and flexible cords was used by the consumer. Also it should be noted that with the continuing increase in the use of electrical equipment in the home, loads are tending to rise, and this is particularly so in kitchens. It is recommended that a separate ring circuit be installed for the kitchen area to handle the loads arising from electric kettles, dish washers, washing machines and other equipments that are likely to be used. In deciding the number of arrangements of ring circuits to be provided, the designer must in any case consider the loads likely to arise so that an appropriate number of circuits are installed.

FIGURE 4.1 Kitchens often require a separate ring circuit to cater for the expected loads. Dish washers, clothes washers and dryers are often encountered in addition to the equipment shown here.

In larger commercial or industrial installations, it is particularly necessary for the division of circuits to be considered carefully. Following the assessment of general characteristics described in Chapter 2, a knowledge of the use of the installation and the nature of the processes which are to be undertaken will guide the designer in the choice of circuit division. In many commercial or industrial installations main and sub-main cables will be needed to supply distribution boards feeding a range of final circuits.

Designed Circuit Current [IEE Regulation 311]

Having decided on the configuration and number of circuits to be used in the installation, the designer next needs to determine the design current for each circuit. Sometimes this can be quite straightforward, and in the case of circuits feeding fixed equipment such as water heaters, no particular problems arise. With circuits feeding socket outlets, the design current will be the same as the rating of the protective device since the designer has no control over the load a user may place on the socket-outlet circuits.

One of the types of installation most difficult to assess is the industrial unit, commonly found on factory estates. In the case where the unit is new or unoccupied, there may be difficulty in determining its future use. It will then be necessary for the electrical designer to install capacity which is judged to be appropriate, informing the person ordering the work of the electrical arrangements provided. This situation is used in the worked example of Chapter 6.

Diversity has already been described previously in this chapter. In assessing the current demand of circuits, allowance for diversity is permitted. It must be emphasised that care should be taken in its use in any particular installation. The designer needs to consider all the information available about the use of the building, processes to be used, etc., and will often be required to apply a degree of experience in arriving at the diversity figure which is appropriate, along with an allowance for future growth.

To summarise, the following steps are required to determine the designed circuit current:

1. Calculate the total installed load to be connected,
2. Calculate the assumed maximum demand after assessing diversity in the light of experience and information obtained,
3. Add an appropriate allowance for future growth.

Nominal Current of Protective Device [IEE Regulation 432]

The nominal current of the protective device is based upon the designed circuit current as calculated above. As allowance has already been made for possible additional future load, the rating of the protective device (i.e. fuse or circuit breaker) should be chosen close to this value. IEE Regulation 433.1.1 demands that the nominal current of the protective device is not less than the design current of the circuit, and choosing a rating as close as possible to it should be the aim. A larger fuse or circuit breaker may not cost any more, but if one of a higher rating is installed it may be necessary to install a larger cable if the protective device is providing overload protection.

The chosen setting current of the protective devices will also need to be selected to avoid unintentional tripping due to the peak current values of the loads [IEE Regulation 533.2.1], for instance circuits containing discharged lighting which may be switched in and out of circuit where the inrush currents may peak above the design current (I_b) of the circuit.

Current-carrying Capacity and Voltage Drop for Cables [IEE Regulation Appendix 4]

Appendix 4 of the IEE Wiring Regulations gives guidance and data on selecting cables. This includes influences such as circuit parameters (ambient temperature, soil thermal resistivity and grouping), the relationship between current-carrying capacities to other factors (such as design current and the setting current of the CPD), overload protection and how the cable size is to be determined. Voltage drop values and the method of installation are also covered.

The data in IEE Appendix 4 is contained in a number of tables. The tables are grouped by letters A to J, with a number representing a sub-reference. They are referred to as rating factors.

Tables 4A1 and 4A2 categorise the methods of installation of cables, Table 4A3 gives the cable specification, current operating temperature and which table from groups 4D to 4J has the applicable current rating and associated voltage drops. 'B' Table groups, 4B1, 4B2 and 4B3, give rating factors relating to temperature, either ambient or soil resistivity. 'C' group tables, 4C1 to 4C5, relate to cable grouping rating factors.

Table 4A1 is a matrix of cable types and permitted installations methods. The installation methods are shown in Table 4A2 where 68 different types are detailed, although there are seven different types of reference methods (A to G).

As mentioned, the current-carrying capacity of the cables differs considerably according to the method of installation which is to be used. This needs to be borne in mind so that the appropriate columns are used when referring to Tables 4D to 4J. Extracts from IEE Tables 4B1, 4D2 and 4D4 are reproduced in Appendix A of this book.

The size of cable to be used for any particular circuit will depend upon a number of factors. Where these factors differ from factors used to tabulate the current-carrying capacities within Appendix 4, a number of rating factors (or corrections factors as they were known) are applied, which are shown within Table 4.3.

A worked example illustrating the various points is given later in Chapter 6, but application of each is detailed below.

Ambient Temperature

The tables giving the current-carrying capacity of cables in the IEE Wiring Regulations are based on an ambient temperature of 30°C in air, this being the ambient temperature used for the United Kingdom, and 20°C for cable in the ground either buried directly or in ducts.

Therefore two tables are provided in the IEE Wiring Regulations giving rating factors to be applied when the ambient temperature deviates from 30°C in free air [IEE Regulation Table 4B1] and 20°C where the cables are buried either directly in the ground or in an underground conduit system [IEE Regulation Table 4B2].

TABLE 4.3 Symbols Used for Rating Factors within the IEE Wiring Regulations

Rating factor	Symbol
Ambient temperature	C_a
Grouping	C_g
Thermal insulation	C_i
Type of protective device or installation condition	C_c
Operating temperature of the conductor (used for voltage drop)	C_t

The ambient temperature is the temperature of the immediate surroundings of the equipment and cables before the temperature of the equipment or cables contributes to the temperature rise. If the ambient temperature is going to be below 30°C, for example in a chilled room, the current-carrying capacity of the conductors can be increased and factors are given in the tables for ambient temperatures of 25°C.

It may be necessary to measure the ambient temperature, for instance, in roof voids or suspended floors where heating pipes are installed since the ambient temperature will be higher in such confined spaces. When measuring such temperatures it is essential that the measurements are not taken close to the items generating the heat, including any cables installed. In checking the ambient temperature for existing cables the measurement should be taken at least 0.5 m from the cables in the horizontal plane and about 150mm below the cables.

The rating factor is applied as a divisor. For example, if a non-armoured multi-core PVC (70°C thermoplastic) cable is installed on a surface in a boiler room where the ambient temperature is 60°C and the circuit is provided with overload protection by a 20A BS88 fuse, what current-carrying capacity must the cable have?

From IEE Table 4B1 for 70°C thermoplastic cable at 55°C the factor is 0.61, therefore,

$$I_t = \frac{I_n}{C_a} = \frac{20A}{0.61} = 32.79A$$

where I_t is the tabulated current-carrying capacity of the conductor being protected and I_n is the rating or setting current of the protective device.

Group Rating Factor [IEE Regulation 523.4]

Before circuits are checked for voltage drop, protection against fault currents and fault protection, it is better to first size the cables. If the cables are grouped with other cables they have to be de-rated. The amount they have to be de-rated is dependent upon how they are installed.

Again, the rating factor is used as a divisor and is divided into the rating of the protective device (I_n) or the design current (I_b) depending upon whether the circuit is protected against overload or not.

Where the cables are spaced twice the diameter of the larger cable apart, no de-rating for grouping is required or if the spacing between adjacent conductors is only one cable diameter then the de-rating factor is reduced. The de-rating factors for grouping will be found in IEE Tables 4C1–4C5 in Appendix 4 and which table is to be used depends on a number of different factors such as the installation method and type of cable.

As an example of how to use these tables, assume that three PVC twin & earth cables are to be bunched together and clipped direct on a surface. In the

IEE Regulations, this is installation method 20, reference method C, and assume they have to be protected against overload by a 20A MCB.

From IEE Table 4C1 the correction factor for three cables bunched together is 0.79. The calculated current-carrying capacity required for the cable (I_t) will therefore be:

$$I_t = \frac{I_n}{C_g} = \frac{20A}{0.79} = 25.32A$$

Using the same example let it be assumed that the cables are supplying fixed resistive loads and that the design current I_b (i.e. full load current) is 17A. What will the current-carrying capacity of the conductors have to be?

There is a difference between a circuit being protected against overload and a circuit that only requires short-circuit protection. In the latter case:

$$I_t = \frac{I_b}{C_g} = \frac{17A}{0.79} = 21.52A$$

Thus when the circuit is only being protected against short-circuit current the current-carrying capacity calculated for the cable size is less, it being based on the circuit's design current instead of the protective device rating.

There are also quite a number of additional notes and provisos that need consideration. These relate to issues such as groups which contain cables that are lightly loaded, of different cables sizes, types or arrangements (e.g. two or three cores, a mixture of single and three-phase circuits etc.) or combination of all or some of these things. In these cases, additional calculations may be required.

Thermal Insulation

Where conductors are in contact with thermal insulation, the thermal insulation reduces the rate of flow of heat from the conductors, thus raising the conductor's temperature. This means that the current-carrying capacity of the conductor has to be reduced to compensate for the reduction in heat loss.

Where cables are to be installed in areas where thermal insulation is likely to be installed in the future, they should be installed in such a position that they will not come into contact with the insulation [IEE Regulation 523.7]. If this is not practicable then the cross-sectional area (CSA) of the conductor has to be increased. The amount the conductor is increased in size is dependent upon the manner in which it is in contact with the insulation.

The current-carrying capacities of cables that are in contact with thermal insulation and on one side with a thermally conductive surface are given in Appendix 4 of the IEE Wiring Regulations. The reference method given to be used in the tables from IEE Table 4A2 is method A.

Where cables are totally enclosed in thermal insulation they have to be de-rated, The Wiring Regulations give the symbol C_i for this. Where the conductors are totally enclosed for 500mm or more, the de-rating factor is 0.5 the values given for reference method C. Where the cables are totally enclosed for distances less than 500mm then the de-rating factor is obtained from Table 52.2 which gives the factors for 50mm (0.88), 100mm (0.78), 200mm (0.63) and 400mm (0.51), for values in between these it would be a matter of inter-polation to give an approximate factor.

Example: A twin & earth (or 70°C thermoplastic insulated and sheathed flat cables with protective conductor) cable is to be installed in a house to a 6kW shower, the cable will be installed along the top of the timber joists in the ceiling clear of any insulation but will be totally enclosed in insulation for 100mm where it passes through the ceiling to the shower unit. To calculate the current-carrying capacity required for the cable:

The connected load is 6000W divided by 230V = 26A.

Since the shower is a fixed resistive load the protective device is only providing fault current protection so that the cable is sized according to the design current I_b (full load current).

From IEE Table 52A the de-rating factor for totally enclosed cables for a distance of 100mm is 0.78; therefore,

$$I_t = \frac{I_b}{C_i} = \frac{26A}{0.78} = 33.3A$$

Overload Protection

Where the protective device is providing overload protection the conductors of the circuit are sized to the nominal rating (I_n) of the protective device. However, it is a requirement of the IEE Wiring Regulations that the current required to operate the protective device (I_2) must not exceed 1.45 times the current-carrying capacity of the circuit conductors (I_z). A correction factor has there-fore to be applied when the protective device does not disconnect the circuit within 1.45 times its rating. This correction factor is generally applied when using a semi-enclosed fuse to BS3036, and shown in the following example:

$$\text{Fusing factor} = \frac{\text{Current required to operate the device } (I_2)}{\text{Nominal rating of the protective device } (I_n)}$$

For a rewirable fuse the fusing factor is 2, therefore

$$2 = \frac{I_2}{I_n} \quad \text{therefore } 2I_n = I_2$$

Substituting for I_2 as stated above into the conditions of IEE Regulation 433.1.1 (iii) as a formula:

$$I_2 < 1.45I_z \quad \text{becomes } 2I_n < 1.45I_z$$

where I_z is the tabulated current-carrying capacity of the conductor being protected. From the formula above $I_n = 0.725\,I_z$ which is more conveniently expressed in the form:

$$I_z = \frac{I_n}{0.725}$$

this means that where semi-enclosed fuses are used for overload protection and only when they are used for overload protection – the de-rating factor 0.725 is applied as a divisor to the rating of the fuse (I_n) to give the current-carrying capacity required for the conductors (I_t). A similar calculation would be necessary for any other type of protective device which had a fusing factor greater than 1.45.

Where, due to the characteristics of the load, a conductor is not likely to carry an overload, overload protection does not need to be provided [IEE Regulation 433.3.1 [ii]]. This means that with a fixed resistive load the protective device is only required to provide protection against fault current. For example, in the case of an immersion heater the load is resistive and an overload is unlikely since it has a fixed load. Under these circumstances the protective device is only providing fault current protection. This means that the conductors can be sized to the design current of the circuit instead of the nominal rating of the protective device.

Example: If a 3kW immersion heater is to be installed on a 230V supply, what is the minimum current-carrying capacity required for the twin & earth (or 70°C thermoplastic insulated and sheathed flat cables with protective conductor) cable and what size of BS 1361 fuse will be required if no other de-rating factors apply?

$$\text{Current taken by immersion heater} = \frac{3000W}{230V} = 13A$$

The fuse size required will be 15A, but as it is only providing fault current protection since the load is a fixed resistive load, I_t for the cable is 13A.

By inspection of Table 4D5 (clipped direct, reference method C) in IEE Appendix 4 it will be found that a 1.0mm^2 cable will be suitable, provided it is not in contact with thermal insulation.

Naturally the circuit must also satisfy the same limitations, placed on it by the IEE Wiring Regulations, as other circuits. As the fuse is only providing fault current protection a calculation will be required to check that the conductors are protected against fault current.

Installation Conditions

The specific installation factor must also be taken in to account as a rating factor, for example where cables are buried in the ground and the surrounding soil resistivity is higher than that stated in the IEE Regulations (2.5Km/W) then an appropriate reduction in the current-carrying capacity must be made. These factors can be found in Table 4B3 of Appendix 4 of the IEE Regulations.

As the C_c rating factor applies for both the type of protective device and the installation conditions, then where they both apply, the composite rating factor is found by simply multiplying the factors together. For example, a cable that is protected by a BS 3036 fuse and is also buried direct in the ground with a thermal resistivity of 3Km/W would equate to a rating factor C_c of 0.6525 (0.725 × 0.9).

Note: where the soil resistivity is unknown, the rating factor is taken as 0.9.

Multiple Rating Factors

So far each of the rating factors has been treated separately, but in practice several of the factors can affect the same circuit. The factors are, however, only applied to that portion of the circuit which they affect. Where each factor is affecting a different part of the circuit the designer can select the factor which will affect the circuit most and size the conductors by using that factor. In other situations a combination of the factors may affect the same portion of the circuit. It therefore follows that a common formula can be remembered that covers all the factors.

This will take the form:

$$I_t = \frac{I_n}{C_a \times C_g \times C_i \times C_t \times C_c}$$

when overload and short-circuit protection is being provided and,

$$I_t = \frac{I_b}{C_a \times C_g \times C_i \times C_t \times C_c}$$ when only short-circuit protection is provided.

In practice the designer will try and avoid those areas where a de-rating factor is applicable, since the application of several factors at the same time would make the cable very large such that, in final circuits, they would not fit into accessory terminations but also be uneconomical.

13A Socket Outlets

There is one combination that has not yet been discussed and that concerns 13A socket-outlet circuits [IEE Regulation 433.1.5]. Standard circuit arrangements are given in Appendix 15 of the IEE Regulations. First, the total load of a socket-outlet circuit is taken as the rating of the protective device protecting the circuit. Secondly, the user of the socket-outlet circuit should be advised as to the maximum load that can be connected to the circuit so that he can ensure $I_b < I_n$.

Thirdly, irrespective of the size of the protective device, socket-outlet circuits can always be overloaded and diversity cannot be taken into account, as this has already been allowed for in the standard circuit arrangements. No de-rating of the circuit cables for grouping is required where two circuits are grouped together (four live conductors). The cables must be de-rated where the ambient temperature exceeds 30°C. Although not mentioned in the IEE *On-site guide* de-rating has to be carried out if the cables are installed in contact with thermal insulation.

Where more than two radial circuits are grouped together de-rating is worked out as for any other circuit protected against overload. Where more than two ring circuits are grouped together the calculation is different. As far as ring circuits are concerned the amount of current in each leg of the ring is dependent upon the socket-outlet distribution round the ring for instance, if all the load was at the mid-point of the ring then the current distribution would be 50% on each leg. As far as standard circuit arrangements are concerned the distribution is taken as being two-thirds in one leg (0.67). The formula therefore for sizing the ring circuit conductors is:

$$I_t = \frac{I_n \times 0.67}{C_a \times C_g \times C_i \times C_t}$$

No diversity in the circuit is allowed since it has already been taken into account. The circuit does not have to be de-rated when a semi-enclosed fuse is used as the protective device.

Relaxation to Grouping Factors

Now that all the types of rating factors have been considered, consideration can be given to some relaxations for cables or circuits grouped together. These relaxations are given in Appendix 4 of the Wiring Regulations and are illustrated below.

Grouped Cables Not Subject to Simultaneous Overload

Where it can be guaranteed that not more than one circuit or cable in the group can be overloaded at any one time, i.e. not subject to simultaneous overload, then the following formulae can be used. Two calculations are required and the calculation that gives the largest I_t is the one used to select the cable size.

Calculation when the protective device is a fuse to BS 88 or BS 1361 or an MCB to BS EN 60898.

$$I_{t1} \geq \frac{I_b}{C_a \times C_g \times C_i} \tag{1}$$

$$I_{t2} \geq \frac{1}{C_a \times C_i} \sqrt{I_n^2 - 0.48 I_b^2 \left[\frac{1 - C_g^2}{C_g^2}\right]} \tag{2}$$

Whichever is the larger, I_{t1} or I_{t2}, is the one that is used to size the cable.

Calculation when the overload protective device is a semi-enclosed fuse

Again where it can be guaranteed that not more than one circuit or cable can be overloaded at any one time, and the protective device is a semi-enclosed fuse providing overload protection the following formulae are used:

$$I_{t1} \geq \frac{I_b}{C_a \times C_g \times C_i \times C_c} \tag{3}$$

$$I_{t2} \geq \frac{1}{C_a \times C_i} \sqrt{1.9I_n^2 - 0.48I_b^2 \left[\frac{1 - C_g^2}{C_g^2}\right]} \tag{4}$$

Again the larger of I_{t1} or I_{t2} is chosen to size the cable.

Care is needed when deciding whether to use this formula to size cables. Where there are a number of socket-outlet circuits grouped together, simultaneous overloading could occur since the designer has no control over what the user of the circuit plugs into it. For example, the use of the ring circuits may only be for PCs, printers, and other accessories etc., but if there is a failure of the heating system then staff may plug in electric heaters to enable work to continue. In any event where protective devices are providing overload protection it is difficult to ensure that only one of the circuits will have an overload occurring at any one time.

Lightly Loaded Conductors

Another relaxation concerns lightly loaded conductors. In this instance if a cable cannot carry more than 30% of its grouped rating, it can be ignored when counting the remaining cables in the group. This relaxation can be very useful where large numbers of control cables are installed in trunking or as armoured cables bunched with power cables.

Using the symbol I_Z for the current-carrying capacity of the cables as given in the tables in IEE Wiring Regulations Appendix 4, the following procedure can be used. Two calculations are required. First the size of the lightly loaded cable has to be determined from the formula previously given, i.e. $I_t = I_n/C_g$ where C_g is the grouping factor for all the cables in the group and I_n is the protective device size for the lightly loaded cables. Having sized the lightly loaded cable the current-carrying capacity given in the table I_Z is noted.

A test is now made by a second calculation to see whether it will be carrying more than 30% of its grouped rating.

$$\text{Grouped current-carrying capacity } \% = \frac{100 \times I_b}{I_Z \times C_g}$$

If the result of this calculation is 30% or less then the cable can be ignored when counting the rest of the cables grouped together. It is important to remember that the size of cable worked out in the first calculation for the lightly loaded cable must be used even if it appears to be too large for the current the cable is carrying.

Example: 70°C Thermoplastic (PVC) insulated single core copper cables are used for 16 single-phase control circuits fused at 2A and are installed in trunking with four power circuits wired in PVC single core cable. If the design current for the control circuits is 1.8A and the power circuits are protected by 32A HRC fuses what size cables can be used?

First determine the minimum size of the control circuit cable:

$$\text{Total number of circuits} = 16 + 4 = 20$$

From Table 4C1 of the regulations, the correction factor for 20 circuits in trunking $= 0.38$.

Calculated current-carrying capacity required for control circuits,

$$I_t = \frac{I_n}{C_g} = \frac{2A}{0.38} = 5.26A$$

From IEE Table 4D1A 1.0mm^2 cable can be used with an I_Z of 13.5A. Now test to see whether the current it is carrying is not more than 30% of its grouped rating.

$$\text{Grouped \%} = \frac{100 \times I_b}{I_Z \times C_g} = \frac{100 \times 1.8}{13.5 \times 0.38} = 35\%$$

This is more than 30% so the cables cannot be excluded from the group in which case the current-carrying capacity required for the power cables is:

$$I_t = \frac{32A}{0.38} = 84.2A$$

From IEE Table 4D1A cable size required for the power cables is 25mm^2.

Using 1.5mm^2 cable for the control circuits (this is the minimum size industrial contractors would normally use).

The first calculation does not need to be repeated. The I_Z for the 1.5mm^2 cable is 17.5A therefore:

$$\text{Grouped \%} = \frac{100 \times I_b}{I_Z \times C_g} = \frac{100 \times 1.8}{17.5 \times 0.38} = 27\%$$

which means that if 1.5mm^2 cables are used for control circuits they can be ignored when counting the number of other cables grouped together.

An alternative way of working out the second calculation once the lightly loaded conductor size has been determined is:

$$I_b < 0.3 \times C_g \times I_Z = 0.3 \times 0.38 \times 17.5 = 1.995A$$

and since I_b is only 1.8A this also proves that the 1.5mm^2 cables can be ignored when determining the CSA of the power cables.

So, the de-rating factor for the power cables now becomes 0.65 for four circuits.

$$I_t = \frac{32A}{0.65} = 49.2A$$

This is a considerable reduction on the previous value of 84.2A. The cable size required now is 10mm^2.

Voltage Drop [IEE Regulation 525 and Appendix 4]

Having determined the size of conductors to install it is now necessary to check that the conductors chosen will comply with the voltage drop constraints demanded by the IEE Wiring Regulations before any further calculations are carried out.

Section 525 of the IEE Regulations covers voltage drop and Regulation 525.1 specifies that under normal operating conditions the voltage at the current-using equipment's terminals must not be less than that specified in the Product Standard for that equipment. Where the input voltage is not specified for equipment by British Standards then the voltage at the equipment must be such as to ensure the safe functioning of the equipment. Where the supply is provided in accordance with the Electricity Safety, Quality & Continuity Regulations 2002, the voltage shall not vary by more than 10% above or 6% below the declared voltage. Regulation 525.3 may be considered satisfied if the voltage drop from the origin of the installation up to the fixed equipment does not exceed the values stated in IEE Appendix 12, which are given in Table 4.4.

When calculating voltage drop, motor starting currents or inrush currents to equipment can be ignored. However, the additional voltage drop caused by these currents has to be considered to ensure the satisfactory starting of equipment. The Regulations do not mention diversity, but diversity can be taken into account when calculating voltage drop, since voltage drop is directly associated with the actual current flowing in conductors.

TABLE 4.4 Maximum Values of Voltage Drop – Extracted from IEE Appendix 12

Origin of supply	Lighting (%)	Other uses (%)
LV installations supplied from a public LV distribution system, i.e. supply provide at LV from the DNO.	3	5
LV installations supplied from a private supply, i.e. consumer provided transformer.	6	8

In general the values for voltage drop allowance given in IEE Appendix 12 are used, but it is up to the designer to ensure that the voltage at the equipment terminals complies with Regulation 525.1 as stated above. The voltage drop must also be split over the sub-main and final circuits, and therefore it is at the designer's discretion how this split is made, it may be a 50/50 split, but if the circuit is feeding for example an external lighting circuit, it may be preferable to allow a greater proportion on the final circuit than the sub-main to ensure an economical design.

All of the current-carrying capacity tables (4D** onwards) in Appendix 4 of the IEE Wiring Regulations have suffix A after the table number. A corresponding voltage drop table is also provided, this having a suffix B after the same table number. A sample current-carrying capacity table is given in Appendix A of this book. The number of this table in the IEE Wiring Regulations is 4D2A. A sample of the voltage drop table is also given, this being shown as Appendix A of this book; the number in the IEE Wiring Regulations for this table is 4D2B.

On inspection of the table in Appendix A, it will be seen that the voltage drops are given in millivolts (mV) per A per metre. These voltage drops are based on the conductor carrying its tabulated current I_t so that the conductor's operating temperature is 70°C. The basic formula for voltage drop is:

$$V_d = \frac{L \times I_b \times mV/A/m}{1000}$$

where L is the length of the circuit, I_b is the full load current and mV/A/m is from the appropriate table; the 1000 converting millivolts into volts (i.e. division by 1000). Two examples will suffice to illustrate how these voltage drop tables are used.

Example 1: A 230V single-phase circuit using a 16mm^2 twin & earth (or 70°C thermoplastic insulated and sheathed flat cables with protective conductors) cable with copper conductors, feeding a single-phase distribution board is 30m long and carries 80A. If the cable is clipped direct and no de-rating factors apply what will the voltage drop be?

From Appendix A of this book the voltage drop mV/A/m for 16mm^2 cable is 2.8.

$$V_d = \frac{30m \times 80A \times 2.8}{1000} = 6.27V$$

which is 2.7% of 230V, and is less than the 5% allowed for a public supply (Table 4.4).

Example 2: A three-phase 400V motor is to be wired in XLPE/SWA/LSF armoured cable having copper conductors. The full load current of the motor is 41A and the length of the circuit is 20m. If the three-phase voltage drop up to the distribution board is 6V what size cable can be used to satisfy voltage drop?

If the normal procedure is adopted of picking a cable and working out the voltage drop several calculations may have to be made before a suitable cable is selected, however, a short cut can be used to size the cable by rearranging the formula.

$$mV/A/m = \frac{V_d \times 1000}{L \times I_b} = \frac{(20\ V - 6\ V) \times 1000}{20 \times 41A} = 17.07\ mV/A/m$$

Note: The 20V is the voltage drop allowed to achieve a maximum of 5% voltage drop on a 400V supply (as Table 4.4).

Now look in the voltage drop in Appendix A [IEE Regulation 4E4A] for a three-core cable whose mV/A/m is less than 17.07, a 2.5mm^2 cable with a mV/A/m of 16 appears to be satisfactory. However, when using this method, the current-carrying capacity of the conductor must always be checked. In the above example although the 2.5mm^2 conductor is satisfactory for voltage drop it will not carry the current and a 4mm^2 (with an I_Z of 42A – assuming the cable is clipped direct) conductor must be used. This highlights the need for this extra check.

Actual voltage drop is:

$$V_d = \frac{20m \times 41A \times 10}{1000} = 8.2V$$

Add to this the voltage dropped up to the distribution board and the total voltage drop is 14.2V, which is less than the 20V allowed.

It will also be noticed that in voltage drop tables of Appendix 4 for cables up to 16mm^2 only one value is given; this is the value using resistance only since it is considered that the reactance in such small cables has a negligible effect. For sizes larger than 16mm^2 three values are given: resistance (r), reactance (x) and impedance (z). For these cables over 16mm^2 where the power factor of the load is not known, the 'z' values are used but where the power factor is known, then the 'r' and 'x' values can be used to provide a more accurate assessment of the voltage drop.

Similarly where the cables are not carrying a significant amount of their tabulated current-carrying capacity I_t, they are also not operating at their maximum temperature and therefore an allowance can be made for the reduction in cable resistance and the effect this has on voltage drop. The calculations are to find the rating factor C_t but they get a little involved and are beyond the scope of this book.

Ring Circuits

Calculation of the voltage drop in a ring circuit, or any other socket-outlet circuit, can only be approximate since the designer will have no control over what equipment the user plugs into each outlet. An average figure can be worked out based on the entire load being at the mid-point of the ring, or one with the load being spread evenly round it. Initially the case with the entire load at the mid-point is calculated, with half the load flowing in half the ring. If the voltage drop limits are satisfied, there is no need to go into the complication of working out the case with the load spread.

Short-circuit Protection

Having determined the size of the conductors the circuits require, the next stage is to check whether they are protected against short-circuit current. It is then necessary to work out the size of protective conductors and then that the arrangements comply with the requirements for protection against indirect contact.

In the case of short-circuit currents the protective device is only providing protection against thermal and mechanical effects that occur due to a fault. It is therefore unnecessary for the conductors to be sized to the rating of the protective device. The relationship that the design current I_b must not be greater than I_n which in turn must not be greater than I_z is related only to overloads and has nothing to do with short-circuit protection.

Two calculations are required: one to give the maximum short-circuit current to enable switchgear or motor starters to be chosen which have the correct fault rating and the second to determine the minimum short-circuit current in order to check that the protective device will operate in the required disconnection time. Examples of the way in which these are calculated are fully covered in Chapter 6.

Fortunately the calculation for checking that the conductors are protected does not have to be carried out in every circumstance. If conductors have been sized to the overload requirements, this means that the conductor's current-carrying capacity is not less than the rating of the protective device. Provided that the breaking capacity of the protective device is not less than the prospective short-circuit current at the point the protective device is installed, it can be assumed that the conductors on the load side of the protective device are protected against short-circuit current. Any circuit breaker in use must be of the current-limiting type. If there is any doubt as to these conditions, a check shall be made. The same applies where cables connected in parallel are to be used [IEE Regulation 435.1].

Obviously this rule will not apply to motor circuits where the protective device has to be sized to allow for starting currents. Nor will it apply where, as in the example of the immersion heater, IEE Regulation 433.3.1 is used and the conductors are sized to the design current and not the protective device rating.

In these cases a calculation is required to check that the conductors are protected against short-circuit current.

The minimum short-circuit current for single-phase and three-phase four-wire supplies will occur with a fault between phase and neutral and the formula is therefore:

$$I_{pn} = \frac{V_{pn}}{Z_{pn}}$$

where I_{pn} is the prospective short-circuit current between phase and neutral, Z_{pn} is the impedance of the phase and neutral from the source to the end of the circuit being checked and V_{pn} is the phase to neutral voltage.

Where the supply is three-phase three-wire then the minimum short-circuit current occurs between two phases and the formula now becomes:

$$I_{pp} = \frac{V_L}{2Z_p}$$

where I_{pp} is the phase to phase prospective short-circuit current, Z_p is the impedance of one phase only from the source to the end of the circuit being considered and V_L is the line voltage.

Checking that the conductors are protected involves working out how long it will take the protective device to disconnect the circuit and then comparing this time with the maximum time allowed for disconnection by the formula given in Regulation 434.5.2, i.e.

$$t \leq \frac{k^2 S^2}{I^2}$$

where t is the maximum disconnection time allowed, k is a factor dependent upon the type of conductor material, the initial temperature at the start of the fault and the limit temperature of the conductor's insulation, obtained from the same regulation. S is the CSA of the conductor and I the fault current.

Where the actual disconnection time is 0.1 s or less then the actual energy let through the protective device must not exceed $k^2 S^2$ this value shall be as defined by the protective device standard or obtained from the manufacturer.

Since it is the minimum short-circuit current that is needed, the resistance of the live conductors is taken at the temperature they reach with the fault current flowing. This involves a complicated calculation so as a compromise the average is taken of the operating temperature of the conductor and the limit temperature of the conductor's insulation. For example, for standard PVC insulation with a copper conductor the maximum operating temperature allowed is 70°C, the limit temperature for the PVC insulation is 160°C so the average will be 115°C and this is the temperature used to determine the resistance of the conductors.

Earth Fault Protection

This form of overcurrent occurs when there is a phase to earth fault, the rules for which will be found in Chapter 54 in the IEE Wiring Regulations. In this case there are two ways of checking whether the thermal capacity of the protective conductor is satisfactory.

The first is to use IEE Table 54.7 where the minimum CSA of the protective conductor should be the same as the line conductor for line conductors greater or equal to 16mm², for 16mm² to greater or equal to 35mm² conductors a minimum of a 16mm² protective conductor should be used and for line conductors greater than 35mm² then a protective conductor of at least ½ the CSA of the line conductor should be used.

If the protective conductor is made from a material which is different from that of the phase conductor, an additional calculation must be made. The size obtained from the table must be multiplied by k_1/k_2, where k_1 is the k factor for the phase conductor material and k_2 is the k factor for the material used for the protective conductor. However, in accordance with IEE Regulation 543.2.3 protective conductors up to and including 10mm² have to be copper.

Apart from the 1.0mm² size of cable none of the other PVC insulated and PVC sheathed cables containing a protective conductor (for instance, twin & earth cable) comply with IEE Table 54.7 since the CSA of protective conductor is less than the phase conductor. The table can be used with aluminium strip armoured cables and for MICC cables since the CSA of the protective conductor at least complies with IEE Table 54.7. For cables larger than 35mm² the table can give a non-standard cable size, in which case the next larger size of cable is selected. In any event the table does not stop the phase earth loop impedance having to be worked out to enable protection against indirect contact to be checked.

The second method is to check whether the protective conductor is protected against thermal effects by calculation by using the formula given in IEE Regulation 543.1.3, known as the adiabatic equation.

$$S = \frac{\sqrt{I^2 t}}{k}$$

where S is the CSA required for the protective conductor, I *is* the earth fault current, t is the actual disconnection time of the protective device and k is a value from IEE Chapter 54 depending upon the conductor material, type of insulation, initial and final temperatures.

It is just a matter of comparing the actual CSA of the protective conductor with the minimum CSA required by calculation.

The value of I is determined from the formula $I = \dfrac{V_{pn}}{Z_s}$

The value of Z_s is obtained from $Z_E + Z_s$ where Z_E is the phase earth loop impedance external to the circuit being checked and Z_s is the phase earth loop impedance for the actual circuit.

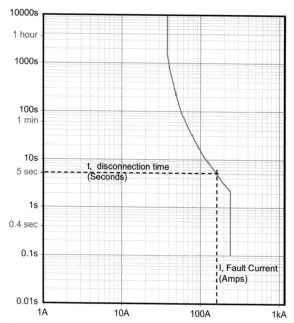

FIGURE 4.2 Finding disconnection time '*t*' using the curve showing characteristic of the protective device (32A Type C MCB).

The disconnection time *t* is then obtained from the protective device characteristic as illustrated in Fig. 4.2. A line is drawn from the current axis for the value of *I* calculated, up to the characteristic for the protective device being used. At the point this line touches the characteristic it is then drawn horizontal to the time axis to give the actual disconnection time for that particular fault current.

The important thing to remember when choosing the value of *k* is the temperature of the conductor at the start of the fault. The protective conductor does not have to be carrying current to have its temperature raised. For instance, a twin & earth cable has the protective conductor in the cable and although it will not be carrying current under normal conditions its temperature will be the same as the live conductors in the cable, and without additional information this is taken as the maximum permitted operating temperature for the cable.

Remember, it is the minimum fault current that is important when proving conductors are protected. So if a calculation has had to be carried out to check that the circuit is protected against short-circuit current and the earth fault current is less than the short-circuit current, it is necessary to recheck that the live conductors are still protected with the earth fault current. Further, since the formula used is the same as for short-circuit protection, if the disconnection time is 0.1 s or less then the manufacturer's let-through energy I_t^2 has to be compared with k^2S^2 in the same way as for short-circuit currents.

One further point that must be remembered is that it is important to take the impedance but not the resistance of conduit, trunking and the armouring of cables.

Automatic Disconnection of the Supply [IEE Regulation 411]

The basic requirement that has to be complied with is given in IEE Regulation 410.1 which implies that the characteristics of the protective device, the earthing arrangements and the impedances of the circuit conductors have to be co-ordinated so that the magnitude and duration of the voltage appearing on simultaneously accessible exposed and extraneous conductive parts during a fault shall not cause danger.

The protective measures available were discussed in Chapter 2, but the following looks at the requirements for fault protection.

Protective Earthing and Equipotential Bonding [IEE Regulation 411.3.1]

It must be clearly understood that the equipotential bonding only allows the same voltage to appear on exposed and extraneous conductive parts within an installation if the fault is outside that installation. The voltage appearing on both the exposed and extraneous conductive parts is the fault current I_f times the impedance of the protective conductor from the source neutral up to the installation's main earth bar, to which the main equipotential bonding conductors are also connected. Where a fault occurs within the installation a voltage will appear on exposed conductive parts and this will be of a higher voltage than that appearing on extraneous conductive parts as illustrated in Fig. 4.3. From the figure it can be seen that the voltage appearing on the exposed conductive part is $I_f(R_2 + R_3)$, whereas the voltage appearing on the

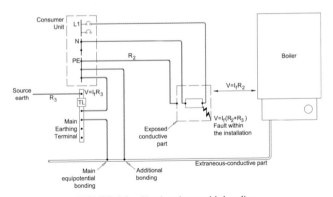

FIGURE 4.3 Shock voltage with bonding.

boiler is $I_f R_3$, the resultant potential difference is $I_f R_2$. This voltage difference may be as high as 150V and is therefore quite dangerous.

Protective equipotential bonding connects the installation's main earthing terminal to the following items via the main protective bonding conductors (in accordance with Chapter 54 of the regulations):

- water installation pipes,
- gas installation pipes,
- other installation pipework and ducting,
- central heating and air conditioning systems,
- exposed metallic structural parts of the building.

Connections may also require to be made to the lightning protection system (in accordance with BS EN 62305) and telecommunication cabling (provided permission is granted by the operator).

Automatic Disconnection in Case of a Fault [IEE Regulation 411.3.2]

The severity of the electric shock a person can receive is governed by three items: first the magnitude of the voltage, secondly the speed of disconnection and thirdly the environmental conditions.

To try and limit the severity of the electric shock when persons are in contact with exposed conductive parts whilst a fault occurs, the IEE Wiring Regulations specify maximum disconnection times depending upon the voltage to earth and the environmental conditions.

The disconnection times for normal environmental conditions – such as those found in carpeted offices which are heated where the occupants are clothed and wearing socks and shoes – are given in IEE Regulation 411.3.2. Where the environmental conditions are not as described above then the disconnection times have to be decreased or other precautions to be taken. Such situations are deemed to be special situations and are covered in Part 7 of the IEE Wiring Regulations.

Generally the disconnection times for a TN system, 230V U_o (open circuit voltage to earth), final circuit is 0.4s up to 32A [IEE Regulation Table 41.1], for circuits over 32A and distribution circuits, the disconnection time can be increased to 5s.

To determine that a circuit will disconnect within the specified time a calculation has to be carried out to determine the earth loop impedance at the end of the circuit. This involves knowing the phase earth loop impedance up to the origin of the circuit referred to as Z_E and the earth loop impedance of the circuit referred to as Z_{inst}. The overall earth loop impedance Z_S can then be determined as follows:

$$Z_S = Z_E + Z_{inst} \ \Omega$$

The value of Z_{inst} represents the impedance of the circuit's phase conductor Z_1 added to the impedance of the circuit's protective conductor Z_2. Tables are available giving $Z_1 + Z_2$ added together for different types of cables, which can be found in Appendix 9 of the IEE On-site Guide. As with earth fault currents it is important to take the impedance of conduit, trunking and the armouring of armoured cables. For circuits up to 35mm^2 impedances Z_1 and Z_2 above are replaced by resistances R_1 and R_2, respectively.

In practice there are two values of Z_S: there is the actual value as calculated or measured and there is the value given in the IEE Wiring Regulations that must not be exceeded. The calculated value is compared with the value allowed to ensure that the disconnection time is not exceeded.

It should be noted that the impedances tabulated within the IEE Regulations will be higher than those within the IEE On-site guide. Figures in the Regulations are design figures at the conductor operating temperature as opposed to those within the On-site guide, which are the measured values at 10°C (i.e. when the testing is carried out).

In a simple installation, such as a house, the value of Z_E will be external to the origin of the installation. In a large commercial building or factory the value of Z_E will be the Z_S of the circuit feeding the incomer of the distribution board and the upstream distribution to the distribution board for the final circuit.

Additional Protection [IEE Regulation 411.3.3]

The requirement for additional protection is covered in Chapter 2. If it is intended to use RCDs as additional protection then this should be in accordance with IEE Regulation 415.1. This regulation permits the use of RCDs provided they have a maximum operating current of 30mA and a disconnection time at five times this current (150mA) causes disconnection within 40ms. Care must be taken to avoid unwanted tripping which could arise when using equipment with high earth leakage currents. Should this problem arise, the solution may be to divide the installation by introducing additional circuits. [IEE Regulations 314.1 (iv) and 531.2.4].

Supplementary equipotential bonding can also be considered as additional protection, and must be applied to all simultaneously accessible exposed conductive parts of fixed equipment and extraneous conductive parts (except those listed in Regulation 410.3.9).

4.3 EARTHING

The object of earthing a consumer's installation is to ensure that all exposed conductive parts and extraneous conductive parts associated with electrical installations are at, or near, earth potential.

Earthing conductors and protective conductors need to satisfy two main requirements, namely to be strong enough to withstand any mechanical damage

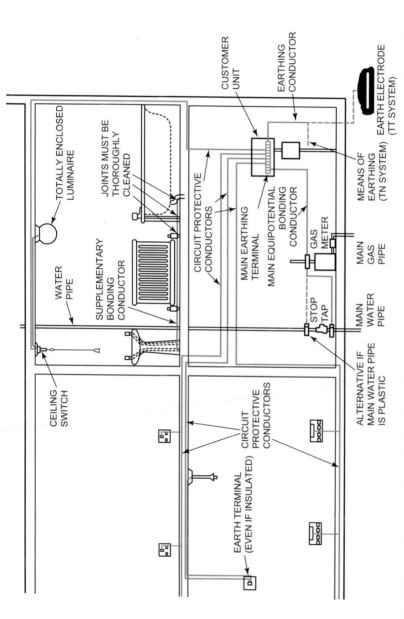

FIGURE 4.4 A diagrammatic representation of a domestic installation showing the main types of protective conductor. The main earthing terminal is normally contained in the consumer unit, and the earthing conductor will be connected to the supply authority's earthing terminal or an earth electrode depending on the system of supply. Note that all the lighting circuits must have a circuit protective conductor even if insulated fittings are used, in which case the CPC is terminated in an earth terminal in the fitting. Special bonding is needed in bathrooms, and ceiling light switches and enclosed luminaires should be used.

Labels in figure:
- CEILING SWITCH
- WATER PIPE
- TOTALLY ENCLOSED LUMINAIRE
- JOINTS MUST BE THOROUGHLY CLEANED
- CUSTOMER UNIT
- EARTHING CONDUCTOR
- SUPPLEMENTARY BONDING CONDUCTOR
- CIRCUIT PROTECTIVE CONDUCTORS
- MAIN EARTHING TERMINAL
- MAIN EQUIPOTENTIAL BONDING CONDUCTOR
- GAS METER
- MAIN GAS PIPE
- MEANS OF EARTHING (TN SYSTEM)
- EARTH ELECTRODE (TT SYSTEM)
- EARTH TERMINAL (EVEN IF INSULATED)
- CIRCUIT PROTECTIVE CONDUCTORS
- STOP TAP
- MAIN WATER PIPE
- ALTERNATIVE IF MAIN WATER PIPE IS PLASTIC

which is likely to occur, and also to be of sufficiently low impedance to meet the need to carry any earth fault currents without danger.

The supply authority connects the neutral point of their transformer to earth, so as to limit the value of the phase voltage to earth. The consumer's earthing system must be so arranged to ensure that in the event of an earth fault of negligible impedance, the fault current shall not be sustained so as to cause danger. The protective devices in the circuit (e.g. fuses or circuit breakers) must operate so as to disconnect the fault within the maximum times specified in the regulations. The protective conductors and earthing system must be arranged so as to ensure that this happens.

IEE Regulations Section 411 and Chapter 54 deal with the design aspects for earthing and the provision of protective conductors, and a number of points require consideration when dealing with this part of the installation design.

The terms used and types of protective conductor are illustrated in Fig. 4.4. The types of protective conductor are shown as follows:

1. The Earthing conductor connects the main earthing terminal with the means of earthing which may be an earth electrode buried in the ground for a TT system, or where a TN system is in use, another means of earthing such as the supply authority terminal.
2. Circuit protective conductors (CPC) are run for each circuit and may comprise a separate conductor, be incorporated in the cable for the circuit concerned or be the metal conduit or cable sheath in, for example, the case of mineral insulated cables. This connects exposed conductive parts of equipment to the main earth terminal.
3. The main equipotential bonding conductor connects the main earthing terminal with the main service metal pipes such as water and gas, and with any exposed building structural steelwork, ventilation ducting etc.
4. Supplementary equipotential bonding conductors are needed in locations where there is increased risk of electric shock [IEE Regulations Part 7 and 415.2] and are used as a form of additional protection. An example would be a bathroom where bonding is required to connect metal parts such as pipes, radiators, accessible building parts baths and shower trays. Bonding is also required in certain special installations such as agricultural sites.

When dealing with the design it is necessary to determine the size of the various protective conductors which are to be used. IEE Regulation 543.1 and Section 544 deal with this aspect and a number of points need to be borne in mind.

The protective conductor must have sufficient strength to protect against mechanical damage. The minimum size where the conductor is not part of a cable is 4mm^2 unless mechanically protected. Thermal considerations are necessary to ensure that when the protective conductor is carrying a fault current, damage to adjacent insulation is avoided. The IEE Regulations give

FIGURE 4.5 A commercial location where a cable riser is constructed from cable basket and carries cables for different types of circuits. Note the equipotential bonding conductors (W.T. Parker Ltd).

two 'standard methods' of determining the cross-sectional size of the protective conductors.

One is by the use of IEE Table 54.7 and the other by using the adiabatic equation calculation. The first standard method involves the use of a look-up table in the IEE Regulations. The second calculation method of determination of protective conductor size is with the use of a formula, and this is given in IEE Regulation 543.1.3 and both of these methods are covered previously in this chapter under 'earth fault protection'.

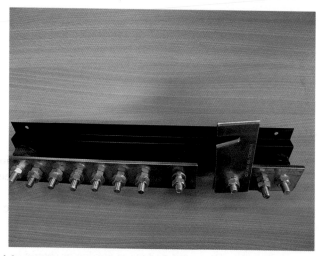

FIGURE 4.6 A main earth bar for use in a commercial installation. Note the link which is normally closed but can be unbolted for testing purposes.

FIGURE 4.7 The main earth terminal in an industrial earthing location. The connections are located where inspection can readily be checked (W.T. Parker Ltd).

Gas and water pipes and other extraneous conductive parts as mentioned above must not be used as an earth electrode of any installation. The consumer's earth terminal may be a connection to the supply undertaking's earth point, if provided by them, otherwise an independent earth electrode must be provided. This consists of buried copper rods, tapes, pipes, or plates etc., as detailed in IEE Regulation 542.2.1.

Most types of armoured multi-core cables rely upon their metal sheathing or armouring to serve as a protective conductor, but it must not be assumed that all multi-core armoured cables have armouring of sufficiently low impedance (especially in long runs of cable) to permit sufficient fault current to flow to operate the protective device.

If the equivalent CSA of the armouring is less than the value required, it may be necessary to increase the sizes of the cable to meet the requirement (although this may be uneconomical) or to provide an additional protective conductor in parallel with the cable or an additional core within the cable.

Where an additional conductor has been provided, the conductor must be sized to carry the earth fault current alone, i.e. the CSA of the armouring cannot be taken into account to reduce the size of the additional protective conductor, as the division of the current which may flow down the conductors cannot be easily predicted.

Metal conduit and trunking are suitable to serve as protective conductors, provided that all joints are properly made, and their conductance is at least equal to the values required in IEE Regulation 543.2.2. Generally a separate productive conductor is provided to guarantee the integrity of the earth.

Flexible or pliable conduit shall not be used as a protective conductor [IEE Regulation 543.2.1], and where final connections are made to motors by means of flexible conduit, a separate circuit protective conductor should be installed within the flexible conduit to bond the motor frame to an earthing terminal on the rigid metal conduit or starter. Additional information may be obtained by reference to British Standard Code of Practice BS 7430 which gives a lot of detailed information on earthing requirements and methods.

In a room containing a fixed bath or shower, Regulation 701.415.2 allows the supplementary equipotential bonding to be omitted if a number of conditions are met, which include the use of RCDs and effective connections between the protective equipotential bonding and all extraneous conductive parts. In case there is any doubt as to these conditions being met, it may still be prudent to provide supplementary equipotential bonding between simultaneously accessible exposed conductive parts of equipment, exposed conductive parts and between extraneous conductive parts effectively meaning that all exposed metalworks, such as pipes, are to be bonded together (Fig. 4.4). Further details can be found in Chapter 7 of this book.

Consideration should be given to bonding in vulnerable situations such as kitchens, laundries, milking parlours, laboratories etc., where persons or animals may be exposed to exceptional risks of electric shock. In these situations, consideration should be given to using residual current circuit breakers.

High Protective Conductor Currents

In cases where an installation is subject to high protective conductor currents, certain requirements are to be implemented. This situation can arise due to the equipment being served and this may include information technology equipment containing switchmode power supplies (SMPS), electronic ballasts in high frequency fluorescent luminaires or variable speed drives (VSD).

There are two aspects of the requirements. In the case of single items of equipment having a protective conductor current between 3.5 and 10mA, these must be either permanently connected or connected via a socket to BS EN 60309-2. For circuits where the combined conductor current will exceed 10mA, a high integrity connection is required.

The regulations detail the provisions to be made depending on the circuit arrangement and the value of conductor current expected to be present. These include specifying the minimum size of protective conductor required, terminating the protective conductors independently, and the provision of an earth monitoring device. The latter must automatically disconnect the circuit if a fault occurs in the protective conductor. The IEE Wiring Regulations should be consulted to determine the appropriate methods to be employed. Other items to consider include the use of RCDs, the effects that high conductor currents may impose and the relevant labelling required at the distribution board.

These requirements were previously treated as a special location within the 16th edition of the Wiring Regulations (Section 607) but are now considered as a common installation and therefore moved into the main body of the regulations (IEE Regulation 543.7).

Protective Multiple Earthing (PME)

Generally new supply systems will be provided as a TN-C-S system employing a PME earthing arrangement where the supply neutral conductor is used to connect the earthing conductor of an installation with earth. The arrangement has some dangers but the main advantage of the system is that any earth fault which occurs automatically becomes a phase to neutral fault, and the consequent low impedance will result in the fast operation of the protective devices.

Multiple earthing of the neutral is a feature of a PME supply and this is employed to ensure that in the event of a broken neutral, dangerous voltages do not occur. High standards of installation are applied by the supply undertaking to reduce the likelihood of an open circuit neutral conductor, and any installation connected to a PME supply must be to the same high standard. If the DNO offers a PME earthing terminal, it may only be used by the consumer if the installation complies with the requirements of the Electricity Safety, Quality and Continuity Regulations 2002. IEE Regulation 411.3.1.2 refers to bonding, and minimum sizes of supplementary bonding conductors are given in IEE Regulation 544.1.1. It is also necessary to look at Regulation 544.2 for sizing the supplementary bonds.

It should be noted that the installation of PME is specifically prohibited in petrol filling stations. The reason being that the exposed conductive parts would have a voltage on them with respect to true earth and parallel earth paths causing high return currents to flow to earth through the petrol station equipment such as dispenser fuel pipes and underground storage tanks must be avoided. Guidance is available in the Energy Institute publication 'Design, construction, modification, maintenance and decommissioning of filling stations'.

4.4 OTHER CONSIDERATIONS

Resistance to Be Used in All Calculations

The Regulation states that an account has to be taken of the increase in resistance of the circuit conductors, which occurs when a fault current flows through them. However, where a protective device complies with the characteristics in IEE Appendix 3 and the earth loop impedance complies with the Z_s values given in the tables in IEE Regulations Part 4, the circuit is deemed to comply with the Regulations. Effectively, the increase in resistance due to increase in temperature can be ignored. The difficulty is determining whether the protective device complies with the characteristics in IEE Appendix 3 since the manufacturers' characteristics are not the same as those given in the Regulations. Additionally, the Z_s values in the tables in IEE Regulations Part 4 will need to be adjusted if the open circuit voltage of the mains supply is different from the 230V on which the tables are based. Where the characteristics do not comply with or are not included in IEE Appendix 3 then the resistance of the conductor is taken at the *average* of the operating temperature of the conductor and the limit temperature of the conductor's insulation. For example, for standard PVC insulation with a copper conductor the maximum operating temperature allowed under normal load conditions is 70°C, the limit temperature for the PVC insulation is 160°C so the average will be:

$$\frac{70 + 160}{2} = 115°C$$

Different types of insulation allow different maximum operating temperatures under normal load conditions, and therefore the average has to be worked out for the type of insulation being used. The average for 90°C thermosetting insulation with a limit temperature of 250°C would be 170°C. Similarly the average will also need to be worked out for protective conductors. This will vary depending upon the type of protective conductor and how it is installed. Where a protective conductor is installed so that it is not in contact with other live conductors its operating temperature will be at ambient temperature which in the United Kingdom is taken as 30°C. The average in this instance will be 95°C.

Where metal sheathed or armoured cables are installed the operating temperature of the sheath is taken as being 60°C, the limit temperature being dependent upon the type of conductor insulation. The tables in IEE Chapter 54 give the assumed initial temperature and limit temperature for different conditions. The average temperature worked out is the temperature used to determine the resistance of the conductors by using the formula:

$$R_{t_2} = R_{20}(1 + \propto (t_2 - t_1)) \; \Omega$$

where t_2 is the final conductor temperature (i.e. 115°C), t_1 is the temperature of the conductor's resistance R_{20} at 20°C and α is the resistance-temperature coefficient for both aluminium and copper. From the simplified formula in BS 6360, α is 0.004. If the values of $t_2 = 115°C$ and $t_1 = 20°C$ are put into the formula it will be found that the resistance T_{t_2} at 115°C is equal to the resistance R_{20} at 20°C multiplied by 1.38.

Reactance has to be taken into account on cables larger than 35mm², but reactance is not affected by temperature so no adjustment for temperature has to be made to the reactance value. Where large cables are used then the impedance of the cable has to be used, not just its resistance, the impedance being obtained from the formula:

$$Z = \sqrt{r^2 + x^2} \; \Omega$$

Multicore Cables in Parallel

It is sometimes desirable to connect two or more multi-core cables in parallel. Potential advantages are in cost or ease of installation as each of the cables used will be smaller than a single cable for the same duty and will have smaller bending radii. However, the cables do have to be de-rated for grouping.

Consider a circuit protected by a 500A CPD, it is assumed that the cables are to be clipped directly to a cable tray. This would require a 240mm^2 four-core Cu XLPE/SWA/LSF cable. If parallel cables were to be installed, two smaller cables could be used. The calculation is shown below. In this example the de-rating factor is 0.88 and

$$I_t = \frac{500A}{0.88 \times 2 \text{ cables}} = 284A$$

where I_t is the calculated current-carrying capacity required for each cable.

An inspection of IEE Table 4D4A shows that two 95mm^2 will be required. If it is possible to install the two cables on the tray so that there is at least one cable diameter between the cables throughout their length – including terminations – the de-rating factor may be reduced further, and possibly the cable size again (although not in this example). These cables could cost less than the original 240mm^2 cable, and a further advantage is that the bending radius of the 95mm^2 cable is much less than the bending radius of the 240mm^2 cable. Where there are confined spaces for the installations of large cables the reduced bending radius may well prove to be of a very considerable advantage, as well as the fact that a 95mm^2 cable is more manageable in terms of both ease of installation and weight.

FIGURE 4.8   The use of cables connected in parallel ease the installation task and can often result in a more economical scheme. This view shows a distribution circuit employing parallel multi-core 415V cables, run on cable ladder, feeding a distribution board. The cables are 120mm^2 multi-core PVC insulated SWA copper. Had parallel cables not been used, 300mm^2 cable would have been needed (W.T. Parker Ltd).

There is no reason why two or more cables should not be connected in parallel but it is important to remember that IEE Regulations require that measures are taken to ensure that the load current is shared equally between them. This can be achieved by using conductors of the same material, CSA and length. Additionally, there must be no branch circuits throughout their length. It must also be remembered that socket-outlet ring circuits are not conductors in parallel and the larger size of the conductor, the less the current is carried per square millimetre of CSA.

For instance a three-core armoured 95mm^2 cable carries 3.2A per mm^2 whereas a 400mm^2 three-core cable is rated to carry only 1.82A per mm^2, therefore nearly half of the conductor of the larger cable performs no useful purpose.

In designing circuits with parallel cables, it is necessary to consider the effect of a fault condition in one conductor only. IEE Regulation 434.5.2 must be applied to check that the characteristic of the protective device is such that the temperature rise of the conductors under fault conditions is contained. The protection of the conductors in parallel against overload should also be considered, the application of this can be found in Appendix 10 of the IEE Regulations.

Four-core Cables with Reduced Neutrals

For multi-core cables feeding three-phase circuits it is permitted to use a reduced neutral conductor. This is not common practice but can be employed providing that there is no serious unbalance between phases and provided the cables are not feeding computing equipment or discharge lighting circuits where significant harmonic currents are likely to occur [IEE Regulation 524.3]. Where the conditions are appropriate, reduced neutrals may also be used when single core cables are installed in conduit or trunking on three-phase circuits.

In cases where the harmonic content of the line conductors may mean that the current-carrying capacity of the neutral conductor may be exceeded, it will be necessary to provide overcurrent detection in line with IEE Regulation 431.2.3.

When feeding panels that control three-phase motors, it is very often satisfactory to install a three-core cable, and should 230V be required at the panel for control circuits this can be provided via a small transformer. This may be cheaper than providing an additional core in a heavy cable but limits the future use of the cable.

Power Factor

Power factor is an inherent feature for example, with the installation of induction motors. The power factor of an induction motor may be as low as 0.6

which means that only 60% of the current is doing useful work. For average machines a power factor of 0.8 lagging is the general rule.

It is advisable, therefore, to understand what power factor means and how it can be measured and improved. In a practical book of this kind no attempt will be made to give in technical terms the theory of power factor, but perhaps a rudimentary idea can be conveyed.

In an inductive circuit, such as exists in the case of an induction motor, the power in the circuit is equal to the instantaneous value of the voltage multiplied by current in amperes, the product being in Watts. A wattmeter, kilowattmeter or kilowatt-hour meter, if placed in circuit, will register these instantaneous values, multiply them, and give a reading in Watts, kilowatts or kilowatt-hours.

Actually in an a.c. circuit the voltage, and therefore the current, varies from zero to maximum and maximum to zero with every cycle.

In an inductive a.c. circuit the current lags behind the voltage. For example, if the normal voltage is 400V and the current is 50A, when the voltage reaches 400V the current may have only reached 30A and by the time the current has risen to 50A the voltage would have fallen to 240V. In either case the total Watts would be 12,000 and not 20,000 as would be in the case of a non-inductive circuit. If a separate ammeter and voltmeter were placed in this circuit the voltmeter would give a steady reading of the nominal 400V and the ammeter would read the nominal 50A, the product of which would be 20,000VA.

A wattmeter in the same circuit would, as explained, multiply the instantaneous values of voltage and current and the product in this case would be 12,000W.

Figure 4.9 shows the components of the power triangle, which are

Component	Symbol	Formula	Unit
Active power	P	VI cos θ	Watts (W)
Reactive power	Q	VI sin θ	Vars (VAr)
Apparent power	S	VI	Volt-amperes (VA)
Power factor	θ	P/S	

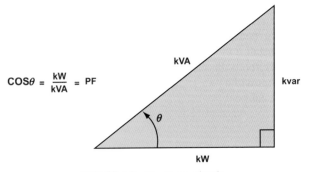

$$COS\theta = \frac{kW}{kVA} = PF$$

FIGURE 4.9   The power triangle.

The relationship between these components of the power triangle is shown in the following formula:

$$P = VI \cos \theta \text{ or } P = \sqrt{3} \, VI \cos \theta \quad \text{for three-phase circuits}$$

and to find the power factor of the circuit in the example above the power (W) is divided by apparent power (VI), i.e.

$$\frac{P}{VI} = \cos \theta = \frac{12,000}{20,000} = 0.6\text{PF}$$

Reactive Power

It will be seen that a portion of the current is not doing useful work, and this is called reactive power or wattless current. Although this current is doing no useful work, it is flowing in the distributors' cables and also in the cables throughout the installation.

As already explained, the kilowatt-hour meter does not register this reactive power and, therefore, when current is charged for on the basis of units consumed the distributor is not paid for this current.

To obtain the true value, either the power factor must be improved, or an additional meter is inserted which records the maximum kilovolt-amperes (kVA) used during a given period. Charges would then be based upon this maximum figure.

Consumers who switch on very heavy loads for short periods are penalised, and thus are encouraged to keep their maximum demand down to reasonable limits.

In these circumstances it will be a paying proposition to take steps to improve the power factor, as this will not only reduce the maximum kVA demand considerably, but also unload the cables feeding the installation.

Improving the Power Factor

This is achieved by the introduction of capacitors, which are generally installed at the main switchboard as part of a Power Factor Correction (PFC) unit. The capacitors introduce capacitance into the circuit which counteracts the lagging power factor caused by inductive loads, the effect of this can be seen in Fig. 4.10.

This can also sometimes be accomplished by means of a synchronous motor, but generally it is advisable to install capacitors. The capacitor only corrects the power factor between the point at which it is inserted and the supply undertaking's generating plant, and therefore from a commercial point of view so long as it is fitted on the consumer's side of the kilovolt-amperes meter its purpose is served. If, however, it is desired to reduce the current in the consumer's cables, then it is advisable to fit the capacitor as near to the load with

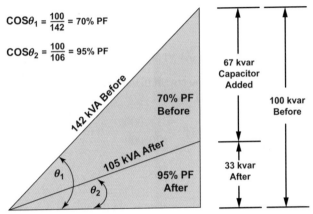

$COS\theta_1 = \frac{100}{142} = 70\%$ PF

$COS\theta_2 = \frac{100}{106} = 95\%$ PF

FIGURE 4.10   By introducing 67 kVA of power factor correction, the PF is improved from 0.7 to 0.95 lagging.

the poor power factor as possible. This means that sometimes there will be a PFC unit installed at the main switchboard and additional units located at the items of equipment with particularly poor power factor. It is the case that most items of plant and equipment (such as discharge lighting) will generally be manufactured with PFC equipment on board.

Where there may be a high presence of any harmonics consideration will need to be made to the de-tuning of the capacitors to counteract the effects that the harmonics have on the PFC equipment. The capacitor has the effect of bringing the current into step with the voltage, but in the part of the circuit that is not covered by the capacitor they get out of step again. Therefore in the motor or equipment itself the power factor still remains low, and its real efficiency is not improved by the installation of a capacitor.

If maximum demand (MD) is based on kilowatts, any improvement in power factor will not result in a reduction of MD charges; if it is based on kVA, however, a considerable saving in MD charges can often be made. Apart from any financial savings, the installation of capacitors to improve the power factor will reduce the current in switchgear and cables, and this can be of considerable advantage when these happen to be loaded up to their limits.

Sustainability

Sustainability can be defined as 'maintaining a process or state' and be applied to nearly every facet of life, particularly from the point of view of maintaining the earth's resources to ensure that they are consumed at a rate which can be replenished.

Sustainability has become a major issue and the way in which building services are designed, installed and implemented can have a significant effect on the natural resources they consume and the environment. The principles behind a sustainable installation will need to be considered even at the most basic level of installation. Methods, processes and equipment selection can all impact on how 'sustainable' the installation will be.

It is outside the scope of this book to detail every aspect of sustainability. However, when designing and installing any electrical system, consideration must be given to the efficiency, material selection, pollution, waste and lifecycles of the products and systems that are to be used. There are numerous guidelines and environmental assessment methods available which can be used as tools to ensure that the most sustainable methods and process have been employed.

Summary

To recapitulate, the stages needed in determining the design of the distribution and final circuits are the following:

1. Calculate the maximum demand taking into account diversity where appropriate.
2. Determine from the electricity supply company whether a supply can be made available for the maximum demand required. Also obtain the following information:
   (a) The type of supply and frequency, is it single or three-phase four-wire, 50/60Hz.
   (b) The earthing arrangement, i.e. the type of system of which the installation will be part.
   (c) The rating and type of the cut-out (fuse) at the origin of the installation.
   (d) The single-phase prospective short-circuit current at the origin (I_{pn}).
   (e) The three-phase symmetrical short-circuit current if the supply is three-phase (I_p).
   (f) The maximum earth loop impedance Z_E at the origin.
3. Work out the distribution arrangement, trying to place distribution boards near the heaviest loads.
4. Determine the type of protective devices that are going to be used throughout the installation.
5. Determine which circuits are being protected against over-load and short-circuit current and those which are being protected only against short-circuit current.
6. Determine what de-rating factors are applicable to each circuit.
7. Calculate the size of live conductors for each circuit.
8. Calculate the voltage drop for each circuit, checking that it is acceptable.
9. Check to ascertain that the conductors chosen are protected against short-circuit current.
10. Calculate the size of protective conductors to be used throughout the installation.
11. Check to ascertain that the circuits give protection against indirect contact.

12. Size main equipotential bonding conductors and determine items to be bonded.
13. Check to see if there are any special situations and if there are, size supplementary bonding conductors.
14. Determine positions of switches, isolators and emergency stop buttons.

This is a general list and there will be cases where all of the above items will not apply, for example, where IEE Regulation 433.3 allows certain circuits not to be protected against fault current. A typical design example is shown in Chapter 6.

4.5 DESIGN BY COMPUTER

Most designs are carried out using a computer package, such as Amtech, Cymap, Hevacomp or Relux. The main issue with using any computer package is that the quality of the result will depend on the accuracy of the input. If the input information is incorrect, then the output will also be incorrect. The packages can only go so far in determining if there is an error with the input information. The user must have a good indication of what results they are expected to achieve, and so will be able to pick up on errors or unexpected outputs, which will cause the user to investigate the anomaly further.

The advantages of using a computer package over manual calculations are that it allows the user to copy, and use standardised cables data, allows faster input. Current computer programs allow the designer to produce fully networked or single radial cable calculations, changing many parameters conveniently and simply without the need for major reworking of the calculation.

Networking of the design allows for the parameter of the upstream components such as fault levels and earth loop impedances to be transmitted through the system, giving a more accurate calculated result. It also allows certain parameters to be calculated back to the very source of the network and can cover every last element of the design, such as voltage drops. The facility also allows the downstream devices to take into account the upstream devices in respect of the additional loads imposed on the system. It gives instant feedback as to the effect and impact caused by any changes to the network.

The input methods range from simple details, giving approximate results and cable sizes required, to complex network calculations enabling advanced fault current characteristics, based on reactance zero fault sequences.

The general process follows a similar path to that of manual calculation, and will generally consist of the following:

1. Input the known characteristics of the source of supply, transformer or generator. This will include the voltage, earth fault, external earth loop impedance. This information is used to determine the prospective fault levels and external characteristics, and most computer programs also allow

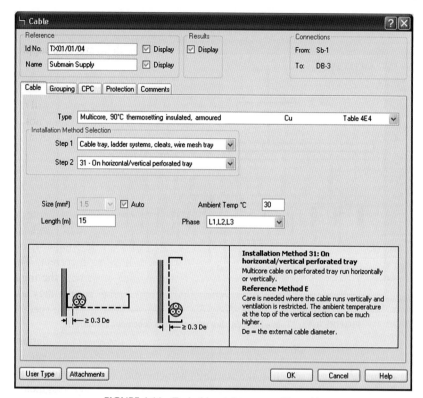

FIGURE 4.11   Typical input data screen (Amtech).

the upstream impedance (i.e. of the High Voltage Network) to be included as this will affect fault current figures.

2. The required outline parameters are entered (Fig. 4.11), such as voltage drop limits, source voltage, phase fault current, earth fault current, ambient temperature and so on.

3. Next the details of the main switchboard, panel boards and main distribution equipment are entered. This information is usually in the form of a schematic diagram (Fig. 4.12), building up the network as information is available. A simplified schematic may have been sketched out previously to determine the general arrangement of the distribution system.

4. The data on the final circuit distribution boards is entered, including their references, function, number of ways, voltage drop allowances in the sub-circuits, and type of protection devices to be used.

5. Details of final equipment such as motor loads, items of fixed equipment and mechanical supplies can be entered along with details of their reference, type (i.e. SPN and TPN), load current, power factor, efficiency, harmonic distortion etc. Also, additional functions may be available in the design

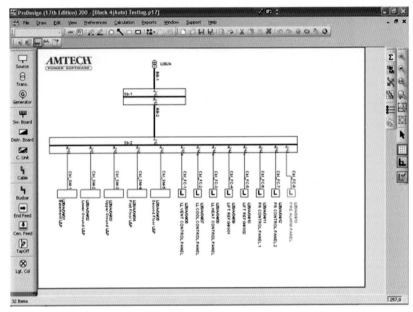

FIGURE 4.12   Typical screen shot of a network schematic (Amtech).

software such as selecting the starting characteristics of large motors, the effect on voltage drop and other constraints on the supply when starting.

6. After all the equipment data has been entered, the cabling is added from point to point on the schematic. This can then be edited to include the details of the length, references, type, installation method, sizes (automatic calculation or a fixed pre-determined size conductor), which phase or phases they are on, ambient temperature, whether cables are installed in parallel, the total number of cables in group and so on. All of which will determine the rating factors, impedances and loads imposed on the network.

7. The computer will then calculate the full system detail and some software packages will perform a logic check of the network to check that everything is connected, perform discrimination checks, select the cable sizes (based on criteria entered) and highlighting any issues that arise. There may also be a facility for adjusting the protection setting of the CPD as required.

Once the system information has been calculated, it is possible to obtain quick results to guide the designer. The process then continues through to very detailed and customised reports such as errors and faults details and fault level analysis. These outputs can be used as a check of the system as it currently stands and can become a deliverable for submission to the client or placed in the design file for further analysis. Most packages also have the option to export the network characteristics and produce the distribution schematic drawings or data ready to be entered into user schedules.

The packages generally have many definable options and settings, which would not necessarily be considered when performing manual calculations. Not all of these affect the design of the system but may provide additional information in the nature of a more complete output. An advantage of using software packages such as these is that if any information is missing, it will be highlighted by the software and prevent calculation continuing until the data is complete.

Chapter 5

Distribution of Supplies in Buildings

This chapter describes some of the points a designer will need to consider when planning an electrical installation.

5.1 INCOMING SUPPLY

In the United Kingdom the electricity distributors, referred to in this text as the District Network Operator (DNO), offer alternative tariffs, and they will always advise consumers as to which is the most favourable tariff after taking into account various factors, such as installed load, type of load, estimated maximum demand and so on. For large industrial installations it may be an advantage for a consumer to purchase electricity at high voltage (HV), although this will entail capital expenditure for HV switchgear and transformers.

Whatever type of installation, whether domestic, commercial or industrial, it is necessary to consult the electricity distributor at an early stage in the designing of an installation, and to make an application for the required size of supply, based on the outcome of the maximum demand assessment made.

The DNO, as the electricity distributor, has discretion as to what supply is provided and when an application is successful, will usually offer a supply in standard denominations to the next available size applied for. A series of information can be obtained from the DNO, such as the prospective supply characteristics and their standard requirements, which will generally be in line with Engineering Recommendations published by the Energy Networks Association (ENA) and any specific DNO requirements which will normally be issued with the acceptance of connection details.

The information detailed will provide the basis for three essential design steps, which:

1. Will provide the supply characteristics (as required by IEE Regulation 132.2) and form the basis of the cable and equipment selection design process,
2. Will determine the electrical supply capacity available and the electrical size of the primary distribution equipment and
3. Will provide the spatial requirements (both physical and operational) and location of the primary distribution equipment.

FIGURE 5.1　Main transformers in a factory building. Incoming supplies are from an 11kV ring main and feed two 800kVA, 11kV to 415V transformers. The distribution board in the sub-station use four-pole 1600A ACBs set at 1200A and a four-pole bus coupler rated at 2000A (W.T. Parker Ltd).

Locating the Incoming Point of Supply (POS)

The DNO can provide the requirements that need to be met when determining the location of their equipment, so co-ordination between the DNO and the professional parties involved (i.e. the client, consultant, architect, structural and civil engineers) needs to take place when determining the incoming POS location. The fire engineer and statutory authorities may also be involved.

General principles include situating the intake position as close as practicable to the incoming cable position, above ground (to reduce risk of flooding), preferably having 24h access, ensuring that adequate space is available to install and operate the equipment safely, securely, and appropriate environmental conditions are maintained.

It will also be preferable to have the main consumer equipment adjacent to the DNO intake position to reduce the length of the service tails to the main switch-panel. Therefore the area should be chosen that is close to boundary and at the centre of the main loads to minimise the length of runs. Biasing the location towards the greatest loads means that the amount of the largest cables is minimised. These will have the greatest losses in Voltage Drop and the most expensive protection, but may also mean that a greater number of smaller sub-mains and or final circuits are required. This could tip the scales the other way in terms of economy, so it becomes a balancing act. Ideally all the greatest loads would be concentrated in the same area with more limited longer runs to smaller loads, but this is often not the case. If the engineer is involved earlier in the project it is sometimes possible to influence the building design and services philosophy.

5.2 MAIN SWITCHGEAR

Every installation, of whatever size, must be controlled by one or more main switches. IEE Regulation 537.1.4 requires that every installation shall be provided with a means of isolation. A linked switch or circuit breaker at the origin shall switch the following conductors of the incoming supply:

1. Both live conductors when the supply is single-phase a.c.
2. All poles of a d.c. supply.
3. All phase conductors in a TP or TP and N, TN-S or TN-C-S system supply.
4. All live conductors in a TP or TP and N, TT or IT system supply.

This must be readily accessible to the consumer and as near as possible to the supply cutouts. The Electricity at Work Regulations 1989 states that 'suitable means ... shall be available for ... cutting off the supply of electrical energy to any electrical equipment'. The type and size of main switchgear will depend upon the type and size of the installation and its total maximum load. Every detached building must have its own means of isolation.

Cables from the supply cutout and the meter to the incoming terminals of the main switch must be provided by the consumer, they should be kept as short as possible, must not exceed 3m, and must be suitably protected against mechanical damage. These cables must have a current rating not less than that of the service fuse and in line with the DNO guidance.

The electricity DNO should be consulted as to their exact requirements as they may vary from district to district. Whatever size of switchgear is installed to control outgoing circuits, the rating of the fuses or the setting of the circuit breaker overloads must be arranged to protect the cable which is connected for

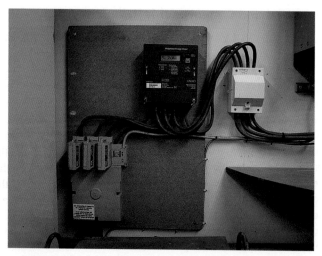

FIGURE 5.2   Typical three-phase commercial intake arrangement.

the time being. If a distribution circuit cable is rated to carry 100A then the setting of the excess-current device must not exceed 100A.

IEE Regulation 430.1 states that every circuit must be protected against overcurrent by a device which will operate automatically and is of adequate breaking capacity. The protective device may, therefore, serve two functions, first to prevent overloading of the circuit, secondly to be capable of interrupting the circuit rapidly and without danger when a short circuit occurs. Although protective devices must be capable of opening the circuit almost instantaneously in the event of a short circuit, they must be sufficiently selective so as not to operate in the event of a temporary overload.

Selection of Switchgear of Suitable Capacity

As has already been pointed out, the main rule which governs all installation work is 'that all apparatus must be sufficient in size and power for the work they are called upon to do'. This applies especially to main switchgear, and it is important to ensure that it is in no danger of being overloaded.

To determine the size required it is necessary to add up the total connected lighting, heating, power and other loads, and then calculate the total maximum current which is likely to flow in the installation. This will depend upon the type of installation, how the premises will be used, whether there are alternative or supplementary means of heating and cooling, and other considerations such as diversity. IEE Regulation 311.1 states that in determining the maximum demand of an installation or parts thereof, diversity may be taken into account. The application of diversity and the calculation of the maximum demand are covered in Chapter 2 of this book.

Large Industrial and Commercial Installations

For loads exceeding 200kVA it is usual for one or more HV transformers to be installed on the consumer's premises. The electricity supplier should be consulted at an early stage to ascertain whether space for a sub-station will be required, and to agree on its position. It is important that it should be sited as near as possible to the heaviest loads so as to avoid long runs of expensive low voltage (LV) cables.

If heavy currents have to be carried for long distances then the size of the cables would have to be increased to avoid excessive voltage drop. This not only increases the cost of the cables, but there would be power losses in the cables for which the consumer will have to pay. It might therefore be advisable to put the sub-station in the centre of where the majority of the load is located.

The sub-station could be provided by the DNO, in which their requirements will need to be sought together with information on the characteristics of the supply which they will be providing. Alternatively, depending on the arrangement of the installation, the sub-station may possibly be provided by the

FIGURE 5.3   A switchboard for use in an industrial premises. The board incorporates over 60 outgoing switches as well as main ACBs, bus-section switches and metering facilities (Pandelco Ltd).

consumer. In this case the only equipment and details required from the DNO will relate to their HV switch and metering point.

Where the consumer provides the sub-station, an option could be the utilisation of a 'Package' sub-station which combines the consumer's HV isolator (where applicable), the step-down transformer and main LV switchboard. This may also incorporate other items of electrical equipment such as the Power Factor Correction (PFC), Electronic Surge Protection (ESP) and control equipment. The main advantages of these packages are that they can be constructed off-site, and have the main cabling between the secondary side of the transformer and the main incoming protective device connected via busbars. This reduces the need for the installation of large cabling and takes up the minimum of space.

When installing LV switchboards for large installations where the supply is derived from a local HV transformer, due consideration must be given to the potential fault current which could develop in the event of a short circuit in or near the switchboard. For example, a 1000kVA 11kV/415V three-phase transformer would probably have a reactance of 4.75%, and therefore the short-circuit power at the switchboard could be as much as 31,000A or 21MVA.

$$\text{PSCC} = \frac{\text{kVA rating} \times 1000}{\sqrt{3} \times U_\text{L}} \times \frac{100}{\%Z}\text{A} = \frac{1000 \times 1000}{\sqrt{3} \times 400} \times \frac{100}{4.75\%}\text{A}$$

$$= 30.388 \text{ kA}$$

There are a number of issues that must be considered when dealing with transformers and the large supplies obtained from them. The specification of

FIGURE 5.4   A sub-station comprising two 1600kVA, 11kV to 415V transformers, incoming and outgoing circuit breakers, fuseswitches controlling outgoing circuits and integral power factor correction (Durham Switchgear Ltd).

the transformer, the type of insulation and cooling, the rating, the vector group, impedance and the protection method all need to be considered.

The greater the impedance of the cables from the secondary of the transformer to the LV switchboard, the lesser will be the potential short-circuit current, and therefore these cables should not be larger than necessary. Most of the LV switchboards are designed to clear faults up to 50kA (for 3s) and would therefore be quite capable of clearing any short-circuit current imposed on a 1000kVA transformer.

If, however, a much larger transformer such as 2MVA (or two 1000kVA transformers connected in parallel) is used then the potential fault current would be as much higher and could exceed the rupturing capacity of standard switchboards. This would entail the installation of a much more expensive switchboard, or special high-reactance transformers, as well as the impact of the increase fault levels on the equipment downstream.

It is usual not to connect transformers having a combined rating exceeding 1500kVA to a standard switchboard and for higher and combined ratings it is usual to split the LV switchboard into two or more separate sections, each section being fed from a single transformer not exceeding say 1500kVA. This method is sometimes applied as it allows greater robustness. Interlocked bus-section switches can be provided to enable one or more sections of the switchboard to be connected to any one transformer in the event of one transformer being out of action, or under circumstances when the load on the two sections of the switchboard is within the capacity of one transformer. Figure 5.5 shows such an arrangement. To ensure that the transformer remaining in service does not become overloaded, it may be necessary to switch off non-essential loads before closing a bus-section switch.

Where an arrangement such as this is provided, consideration should be paid to the use of load shedding by the arrangement of the circuits being supplied to enable non-essential loads to be lost, but essential and life-safety loads to remain. This can sometimes be achieved by the use of automatic control systems which sense the lost of supply and initialise the sequence to enable the supply to be switched over to the live supply while shedding the non-essential supply to maintain the supplies to the loads that are required.

Where applications such as this are justified, it would be normal to see the use of a stand-by supply, maybe in the form of a diesel generator, which will take up the load in the event of a loss of mains. There are a number of considerations associated with this arrangement as well as the load shedding, the load step that the generator will see when it takes up the supply must be considered as too great an initial load imposed on the generator may stall or lock the generator out. In these cases it may be necessary for the control system to gradually reinstate the supplies onto the generator. In addition, the earthing and neutral arrangement as well as the protection method will need careful consideration.

If the supplies are critical, then the change-over from the mains to the stand-by supply and back may need to be achieved without loss of mains at all, in which case the generator and the incoming supply(s) may have to run in parallel for a short period of time, again there are a number of conditions and consideration that are required to be met to enable this to happen, and approval from the DNO will need to be obtained. In most cases similar to this, it may be

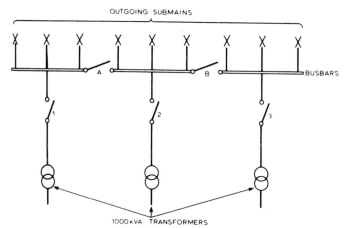

FIGURE 5.5 Arrangement of bus-section switches on LV switchboard (single line diagram). 1, 2 and 3: main switches. A and B: bus-section switches. Bus-section switches A and B are normally open. These are interlocked with main switches 1, 2 and 3. A can only be closed when 1 or 2 is in Off position. B can only be closed when 2 or 3 is in Off position. This enables one transformer to take the load of two sections of the LV switchboard if required.

simpler to provide a clean 'break' from the mains and to provide an alternative method to keep any critical supplies running [such as the use of Uninterruptible Power Supplies (UPS)] while the generator is starting and unable to accept the load.

Main switchgear for industrial and other similar installations, such as commercial buildings, hospitals and schools, will be designed and rated according to the maximum current that is likely to be used at peak periods, and in extreme cases might be as much as 100% of the installed load. For such installations it is usual to provide main switchgear, not only of sufficient size to carry the installed load, but to allow ample margins for future extensions to the load.

Switchboards

A protected type switchboard (Fig. 5.6) is one where all of the conductors are protected by metal or other enclosures. They generally consist of a bespoke metal cubicle panel, or a modular arrangement mounted into a standardised frame, which can be customised by a number of different modules to provide the exact arrangement required for the installation. They usually consist of an incoming section(s) and main switch, busbar sections interconnected to distribute between the outgoing sections and the outgoing sections which can consist of circuit breakers, fuses or even motor starters. The switchboards can be arranged to provide a number of options, including multiple incoming sections, interconnecting busbar and change-over arrangements, integral PFC, ESP, motor control and metering sections. Circuits and conductors are normally segregated within the switchboard and various levels of segregation may be used.

This segregation is referred to as the 'forms of separation', and these are detailed in BS EN 60439-1. There are four main forms of separation, with each of these forms having as many as seven different types. Generally the higher the 'form' the more protected the switchboard is in terms of segregation between live parts, functional units and cables. They all provide different methods of preventing faults occurring on one circuit from transferring to the adjacent circuits. This provides increased protection to persons operating and maintaining the switchboards, although the necessary requirements for isolation and safe working will still need to be adhered to. Higher 'form' switchboards take up more space (and are likely to be more expensive) and so due consideration will need to be paid as to the providing of the correct type of switchboard for the required application.

Electricity at Work Regulation 15 gives other requirements which apply to switchboards. These include such matters as the need for adequate space behind and in front of switchboards; there shall be an even floor free from obstructions, all parts of which have to be handled shall be readily accessible, it must be possible to trace every conductor and to distinguish between these and those of other systems, and all bare conductors must be placed or protected so as to prevent accidental short circuit.

FIGURE 5.6 A protected switchboard with separate lockable compartments to house the incoming and outgoing cabling. Although these can be accessed from the front of the panel, it is still essential to allow space behind the panel to allow subsequent maintenance to be carried out at the rear (Pandelco Ltd).

Some older switchboards used to exist and may occasionally be encountered in old installations. They are referred to as open-type switchboards and the current-carrying parts are exposed on the front of the panels. The type is rarely used, but where they do exist a handrail or barrier must be provided to prevent unintentional or accidental contact with exposed live parts. They must be located in a special switchroom or enclosure and only competent persons may have access to these switchboards.

Busbar chambers which feed two or more circuits must be controlled by a switch, circuit breaker, links or fuses to enable them to be disconnected from the supply to comply with IEE Regulation 131.15.1.

Other Considerations for Selection of Main Switchgear

Earthed neutrals: To comply with IEE Regulation 131.14.2, and Regulation 9 of the Electricity at Work Regulations 1989, no fuse or circuit breaker other

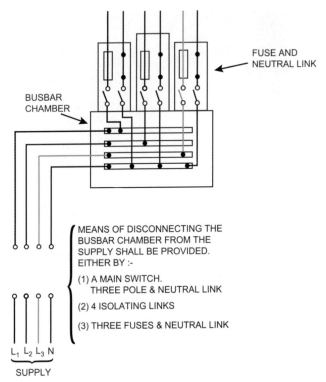

FIGURE 5.7   Isolation of busbar chamber. Busbar chambers must have a means of disconnection from the supply.

than a linked circuit breaker shall be inserted in an earthed neutral conductor, and any linked circuit breaker inserted in an earthed neutral conductor shall be arranged to break all the related phase conductors.

These regulations cover PME supplies and the above rule applies throughout the installation, including two-wire final circuits. This means that no fuses may be inserted in the neutral or common return wire, and the neutral should consist of a bolted solid link, or part of a linked switch which completely disconnects the whole system from the supply. This linked switch must be arranged so that the neutral makes before, and breaks after the phases.

Under certain systems of supply, the star-point of the transformer will require to be earthed, which also forms the neutral point of the system. Where this neutral-earth point occurs will depend on the arrangement and protection requirements of the supply, but it is usually made at either the actual star-point of the transformer or brought out to the main switchboard for connection. Whichever the arrangement, careful consideration will be required and consultation with the DNO.

Power Factor Correction (PFC): This equipment is sometimes provided at the main switchboard and this improves the power factor of the installation.

It is usually in the form of banks of capacitors which automatically switch in and out of circuit to correct power factor of the installation. They are generally arranged in a number of banks, and may also incorporate inductors and de-tuning circuits to counteract the presence of any harmonics that may exist. Chapter 4 gives further details on power factor and power factor correction.

Electronic Surge Protection (ESP): This is quite often provided at main switchboards, as well as at any other parts of the installations that may be susceptible. This is in the form of a unit supplied either from one of the outgoing feeders or directly onto the main busbars of the panel. These are designed to supplement other forms of protection against transient over-voltage.

Transient over-voltages are usually caused by either direct or indirect lightning strikes, or switching events upstream of the incoming supply. When a transient over-voltage occurs, it may affect sensitive electronic equipment either by disruption or by direct damage to a system. Whether protection is required is the subject of risk assessment procedure, although on larger installations the comparably low cost of providing protection may outweigh the possible risk if it was not to be provided.

5.3 FINAL CIRCUIT SWITCHGEAR

Distribution Boards

A distribution board may be defined as 'a unit comprising one or more protective devices against overcurrent and ensuring the distribution of electrical energy to the circuits'. Very often it is necessary to install a cable which is larger than would normally be required, in order to limit voltage drop, and sometimes the main terminals are not of sufficient size to accommodate these larger cables. Therefore distribution boards should be selected with main terminals of sufficient size for these larger cables, although extension boxes may also be utilised to assist with glanding the cables and allow space for accessories such as metering.

Types of Distribution Boards

The main types of distribution boards are (1) those fitted with HRC fuselinks, (2) those fitted with circuit breakers, and (3) Moulded Case Circuit Breaker (MCCB) panel boards. Distribution boards fitted with miniature circuit breakers (MCBs) are more expensive in their first cost, but they have much to commend them, especially as they can incorporate an earth-leakage trip. MCBs are obtainable in ratings from 5A to 63A, all of which are of the same physical size. When assembling or installing the distribution board, care must be taken to ensure that the MCBs are to the correct rating for the cables they protect. Every distribution board must be connected to either a main switchfuse or

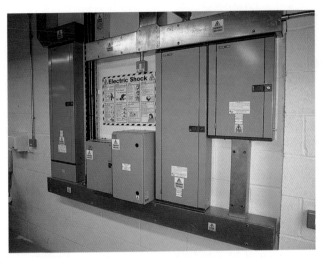

FIGURE 5.8 A distribution board in use in a college premises. The board incorporates three MCB distribution panels and associated switches. The installation is neatly wired in steel trunking and on cable tray.

a separate way on a main distribution board. Every final circuit must be connected to either a switchfuse, or to one way of a distribution board.

Positions of Distribution Boards

As with main switchgear, distribution boards should preferably be sited as near as possible to the centre of the loads they are intended to control. This will minimise the length and cost of final circuit cables, but this must be balanced against the cost of sub-main cables. Other factors which will help to decide the best position of distribution boards are the availability of suitable stanchions or walls, the ease with which circuit wiring can be run to the position chosen, accessibility for replacement of fuselinks, and freedom from dampness and adverse conditions.

Supplies Exceeding 230V a.c.

Where distribution boards are fed from a supply exceeding 230V, feed circuits with a voltage not exceeding 230V, then precautions must be taken to avoid accidental shock at the higher voltage between the terminals of two lower voltage boards.

For example, if one distribution board were fed from the L1 phase of a 415/240V system of supply, and another from the L2 phase, it would be possible for a person to receive a 415V shock if live parts of both boards were touched simultaneously. In the same way it would be possible for a person to receive a 415V shock from a three-phase distribution board, or switchgear.

IEE Regulation 514.10 requires that where the voltage exceeds 230V, a clearly visible warning label must be provided, warning of the maximum voltage which

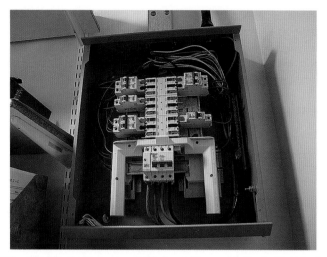

FIGURE 5.9 An MCB distribution board. The board illustrated is fitted with eight single-phase MCBs feeding the final circuits, fed by two of the phases (brown and black). Notices fixed to the outside of the board warn of voltages exceeding 230V.

exists. These warning notices should be fixed on the outside of busbar chambers, distribution boards or switchgear, whenever voltage exceeding 230V exists.

Feeding Distribution Boards

When more than one distribution board is fed from a single distribution circuit, or from a rising busbar trunking, it is advisable to provide local isolation near each distribution board. It is also necessary to provide a local isolator for all distribution boards which are situated remote from the main switchboard, since IEE Regulations 131.15.01 calls for every installation and circuit to be provided with isolation and switching.

If the main or sub-main consists of a rising busbar or insulated cables in metal trunking, it is very often convenient to fit the distribution boards adjacent to the rising trunking, and to control each board with fusible cutouts or a switchfuse.

Circuit Charts and Labelling

IEE Regulation 514.9.1 requires that diagrams, charts or tables shall be provided to indicate the type and composition of each circuit. Details of this requirement are quite comprehensive and are given in IEE Regulation 410.3.3.

Marking Distribution Boards

All distribution boards should be identified by marking them with a letter, a number or both. Suitable prefixes may be L for lighting, S for sockets and P for power for consistency. They should also be marked with the voltage and the type of supply, and if the supply exceeds 250V a DANGER notice must be fixed.

When planning an installation a margin of spare ways should be provided – usually about 20% of the total and this must be matched by an increase in the current-carrying capacity of the distribution cables. Distribution boards are usually provided with a number of 'knockouts' to enable additional conduits or multicore cables to be easily connected in future.

Main Switchgear for Domestic Installations

It is usual to install a domestic consumer unit as the main switchgear, and also as the distribution point in a small or domestic installation. A wide range of makes and types of consumer unit is available. These units usually consist of a main switch of up to 100A capacity, and an associated group of single-pole ways for overcurrent protection of individual circuits. No main fuse is normally used with these units as the supply undertaking's service fuse will often provide the necessary protection of the tails connecting the fuse to the consumer unit. To ensure that this is so, a knowledge of the prospective short-circuit currents is necessary, and the breaking capacity of the devices to be used. This is covered in more detail in Chapter 2 of this book.

Generally the protective devices fitted in the unit will be MCBs, Residual Current Devices (RCDs) or RCBOs. HRC fuses can be used but are less flexible, require the complete fuse to be replaced if operated and may need to be combined with additional equipment, such as RCDs, to meet the requirements of enhanced protection defined by the IEE Regulations. Semi-enclosed fuses may also be present in older installations, but they are not generally installed in new installations, and it is usually possible to find an MCB replacement that is a direct replacement for existing semi-enclosed fuse carriers.

FIGURE 5.10 A 63-A RCCB with a 30-mA tripping current is fitted in a school and protects specific parts of the installation.

Split way consumer units are especially useful where a TT system is in use, as the residual current protection enables the regulations for basic protection to be complied with. It should be noted that there is not necessarily any benefit in providing residual current protection on circuits where it is not strictly necessary as this may introduce nuisance tripping [IEE Regulation 314.1] and provided the installation design is such that the correct disconnection times are obtainable, normal overcurrent protection may suffice.

To take an example, an RCD is needed for any sockets intended for equipment being used outdoors. If this RCD is one in a consumer unit which acts on all the circuits, a fault on one circuit will trip the residual current circuit breaker and disconnect the whole installation. In order to avoid any inconvenience to the users, it would be better therefore to provide the residual current protection only on the circuits which demand it.

5.4 CIRCUIT PROTECTIVE DEVICES (CPDS)

Types of Protection

When selecting the equipment to be utilised for the electrical installation, one of the fundamental issues is the choice of protective device to be used. There are a number of types available and they all have their individual merits. The selection will influence the design criteria, cable sizing and other factors which will need to be considered as part of the design process. The first consideration is the load that is to be protected, whether a main switchboard, sub-main circuit or final circuit. The general types are detailed below.

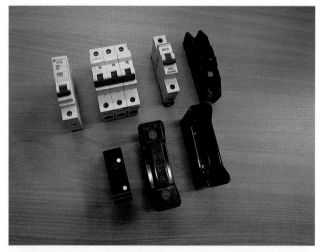

FIGURE 5.11 Overcurrent protective devices. Single- and three-phase MCB (top), a BS 1361 fuse (lower right) and rewirable fuses to BS 3036 (lower left). The use of HRC fuses or MCBs is strongly recommended.

Fuses

HRC fuses: HRC fuses to BS 88 and cartridge fuses to BS 1361 (Fig. 5.12) will give discriminate protection against overcurrents, and will also clear short-circuit currents rapidly and safely up to their rated breaking capacity. They can be used for both sub-main and final circuit distributions.

For this reason HRC fuselinks are designed so that they will withstand as much as five times full load current for a few seconds, by which time the fault will probably be cleared by a final CPD, or local control gear. If main HRC fuses are carefully selected and graded so as to function with discrimination, the final CPD will take care of all normal overloads and short circuits. These main fuses will operate only when the short circuit is in the feeder cable the fuse is protecting, or in the event of the cumulative load of the final circuits exceeding the rating of the main fuses.

Special HRC fuses are sometimes needed for motor circuits to take care of heavy starting currents, and normal overcurrent protection for these circuits is provided in the motor starters.

Rewirable or semi-enclosed fuses made to BS 3036 are mainly confined to domestic installations, and offer a crude method of overcurrent protection in comparison to HRC fuses and MCBs. Their use inevitably means that larger cables are required and the time is not far distant when this type of protection will be a thing of the past.

Circuit Breakers

Circuit breakers are designed to handle safely heavy short-circuit currents in the same manner as HRC fuses.

FIGURE 5.12 A range of small sizes of HRC fuses to BS 88 and BS 1361.

Such circuit breakers have a number of advantages over other types of circuit protection. However, care is needed in selection and maintenance to ensure compliance with Regulation 5 of the Electricity at Work Regulations, 1989, which requires that the arrangements must not give rise to danger, even under overload conditions. If a moulded case circuit breaker has had to clear faults at its full rated breaking capacity, it may need to be replaced to ensure that it can interrupt a fault current safely.

Circuit breakers do have some inherent advantages. In the event of a fault, or overload, all poles are simultaneously disconnected from the supply. Some types of devices are capable of remote operation, for example, by emergency stop buttons, and some have overloads capable of adjustment within pre-determined limits.

There are a number of types of circuit breakers, the type required will depend on a number of factors, but mainly is determined by design current of the circuit they serve, the fault handling capacity and the need for discrimination with the protective device both up- and downstream of the device.

Generally for circuits up to 63A (i.e. final circuits) an MCB would be utilised, for supplies of 63–800A (i.e. sub-main circuits) an MCCB may be used, and for supplies above 800A (i.e. supplies to main switchboards) Air Circuit Breakers (ACBs) would be used, although the ranges of devices do overlap.

Miniature circuit breakers (MCBs): Circuit breakers have characteristics similar to HRC fuses, and they give both overcurrent protection and short-circuit protection. They are normally fitted with a thermal device for overcurrent protection, and a magnetic device for speedy short-circuit protection.

A typical time/current characteristic curve for a 20A MCB is shown in Fig. 5.13 together with the characteristic for a 20A HRC fuse. The lines indicate the disconnection times for the devices when subject to various fault currents. MCBs typically have a range from 1.5A to 125A, with breaking capacities of up to 16kA, but the manufacturers' data should be consulted to determine the rating for a particular device. MCBs can be obtained combined with RCDs, and these can be useful where RCD protection is a requirement.

Residual current circuit breaker: As stated within Chapter 2, rapid disconnection for protection against shock by indirect contact can be achieved by the use of an RCD. A common form of such a device is a residual current circuit breaker. The method of operation is as follows. The currents in both the phase and neutral conductors are passed through the residual current circuit breaker, and in normal operating circumstances the values of the currents in the windings are equal. Because the currents balance, there is no induced current in the trip coil of the device. If an earth fault occurs in the circuit, the phase and neutral currents no longer balance and the residual current which results will cause the operation of the trip coil of the device. This will in turn disconnect the circuit by opening the main contacts.

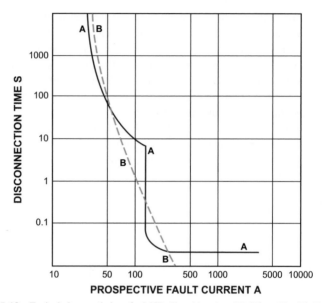

FIGURE 5.13 Typical characteristics of a MCB (line A) and an HBC fuse (line B). Both are for 20A rated devices.

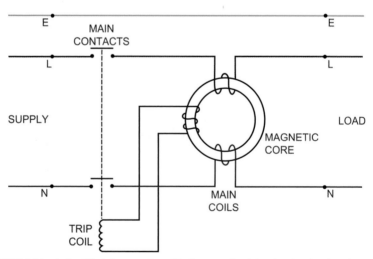

FIGURE 5.14 A simplified diagram of a residual current circuit breaker showing the windings as described in the text.

IEE Regulations call for RCDs to be used to protect any socket which can be expected to be used for supplying outdoor equipment [IEE Regulation 411.3.3] and is preferred for any socket outlets which are part of a TT system [IEE Regulation 411.5.2]. RCDs may also be used if difficulties are experienced in obtaining sufficiently low earth fault loop impedance to obtain a satisfactory disconnection time.

It should be noted that RCDs cannot be used where a PEN (combined protective and neutral) conductor is in use on the load side of the RCD for the simple reason that even in earth fault conditions the currents will balance and there will be no residual current to operate the breaker [IEE Regulation 411.4.4].

Moulded Case Circuit Breaker (MCCB): This works on principles similar to that of MCBs except that they generally use more sophisticated techniques to extinguish the arc such as arc chutes and magnetic blowout coils. They also provide a wider range of protection options, from thermal magnetic to fully electronic relays and are designed to handle much larger currents and fault levels. MCCBs typically have a range from 16A to 1600A, with breaking capacities of 36kA up to 150kA.

Air Circuit Breakers (ACBs): These are the next stage on from MCCBs and use much more sophisticated protection relays to enable the characteristics to be set very accurately. ACBs may use compressed air to blow out the arc, or the contacts are moved rapidly blowing out the arc. ACBs typically have a range from 1000A to 6300A, with breaking capacities of 40kA up to 150kA.

FIGURE 5.15 A view of an MCCB. This device incorporates both bimetallic and magnetic trip mechanisms to open the contacts under overload or short-circuit conditions. The operating toggle has three positions and shows when the breaker has tripped. A range of auxiliary components can be fitted such as undervoltage releases, or control interlocks. These MCCBs can be obtained with breaking capacities up to 150kA.

5.5 CABLING AND DISTRIBUTION

Colour Identification of Cables and Conductors

IEE Regulation 514 lays down the requirement for identification of conductors and Regulation 514.3.2 states that every core of a cable shall be identifiable at its terminations and preferably throughout its length and IEE Table 51 specifies the alphanumeric and colour identification to be used. There are a few exceptions to this and these include concentric conductors, metal sheaths or armouring when used as a protective conductor and bare conductors where permanent identification is not practicable. Table 5.1 summarises the colour requirements and includes extracts from IEE Table 51.

Although colour identification alone is permitted at interfaces in single-phase installations, additional permanent alphanumeric marking is required in two- or three-phase schemes. It will be appreciated that the old phase colour blue must not be confused with the new neutral cable colour which is also blue. The table lays down alphanumeric symbols to be used and, at a three-phase interface, both existing and additional cores shall be marked 'N' for neutral conductors and 'L1', 'L2' or 'L3' for phase conductors. In any installation, whether single- or three-phase, where two different colour standards are present, a warning notice must be affixed at or near distribution boards. This is shown in Fig. 5.16.

CAUTION
This installation has wiring colours to two versions of BS 7671. Great care should be taken before undertaking extension, alteration or repair that all conductors are correctly identified.

FIGURE 5.16 Warning notice required by the IEE Regulations where mixed wiring colours occur in an installation.

Switch wires.   It is usual to run a two-core and cpc cable with cores coloured brown and blue to a switch position, both conductors being phase conductors. In such a case, the blue conductor must be sleeved brown or marked 'L' at the terminations. The same applies to the black and grey cores of three-core cables if used in intermediate or two-way switched circuits.

MI cables.   At the termination of these cables, sleeves or markers shall be fitted so that the cores are identified and comply with IEE Table 51.

Bare conductors.   Where practical, as in the case of busbars, these are to be fitted with sleeves, discs, tapes or painted to comply with IEE Table 51. An exception is made where this would be impractical such as with the sliding contact conductors of gantry cranes, but even then, identification would be possible at the terminations.

Motor circuits.   When wiring to motors, the colours specified in IEE Table 51 should be used right up to the motor terminal box. For slip-ring

TABLE 5.1 Colour Identification of Conductors and Cables (Includes Extracts from IEE Table 51)

Function	Alpha-numeric	Colour (IEE Table 51)	Old fixed wiring colour
Protective conductors		Green and yellow	Green and yellow
Functional earthing conductor		Cream	Cream
a.c. Power circuit (including lighting)			
Phase of single-phase circuit	L	Brown	Red
Phase 1 of three-phase circuit	L1	Brown	Red
Phase 2 of three-phase circuit	L2	Black	Yellow
Phase 3 of three-phase circuit	L3	Grey	Blue
Neutral for single- or three-phase	N	Blue Black circuit	
Two-wire unearthed d.c. circuits			
Positive	L1	Brown	Red
Negative	L2	Grey	Black
Two-wire earthed d.c. circuit			
Positive (of negative earthed) circuit	L1	Brown	Red
Negative (of negative earthed) circuit	M	Blue	Black
Positive (of positive earthed) circuit	M	Blue	Black
Negative (of positive earthed) circuit	L2	Grey	Blue
Three-wire d.c. circuit			
Outer positive of two-wire circuit derived from three-wire system	L1	Brown	Red
Outer negative of two-wire circuit derived from three-wire system	L2	Grey	Red
Positive of three-wire circuit	L1	Brown	Red
Mid wire of three-wire circuit	M	Blue	Black
Negative of three-wire circuit	L2	Grey	Blue
Control circuits, extra-low voltage etc.			
Phase conductor	L	Brown, Black, Red, Orange, Yellow, Violet, Grey, White, Pink or Turquoise	
Neutral or mid wire	N or M	Blue	

motors, the colours for the rotor cables should be the same as those for phase cables, or could be all one colour except blue, green or green and yellow.

For star delta connections between the starter and the motor, use Brown for A1 and A0, Black for B1 and B0 and Grey for C1 and C0. The 1 cables should be marked to distinguish them from the 0 cables.

Distribution Circuits

Distribution circuits (sometimes referred to as sub-mains) are those which connect between a main switchboard, a switch fuse, or a main distribution board to sub-distribution boards. The size of these cables will be determined by the total connected load which they supply, with due consideration for diversity and voltage drop, and the other factors described in Chapter 2.

Distribution circuits may be arranged to feed more than one distribution board if desired. They may be arranged to form a ring circuit, or a radial circuit looping from one distribution board to another, although this is not common practice. Where a distribution circuit feeds more than one distribution board its size must not be reduced when feeding the second or subsequent board, because the cable must have a current rating not less than the fuse or circuit breaker protecting the sub-main [IEE Regulation 433.2.1].

If a fuse or circuit breaker is inserted at the point where a reduction in the size of the cable is proposed, then a reduced size of cable may be used, providing that the protective device is rated to protect the cable it controls.

FIGURE 5.17 A 13-way consumer unit with final circuit MCBs and incorporating a 30mA RCD protective device.

FIGURE 5.18 MCB distribution boards form a convenient way of arranging distribution of supplies. They can be obtained in a range of sizes, and the illustration shows the board with covers removed (W.T. Parker Ltd).

5.6 FINAL CIRCUITS

Design and arrangement of final circuits: Previous chapters dealt with the control and distribution of supply and described the necessary equipment from the incoming supply to the final distribution boards. The planning and arrangement of final circuits, the number of outlets per circuit, overload protection, the method of determining the correct size of cables and similar matters are dealt with in this section, and it is essential that these matters should be fully understood before proceeding with practical installation work.

Definition of a 'final circuit': A final circuit is one which is connected directly to current-using equipment, or to socket outlets for the purpose of feeding such equipment. From this it will be seen that a final circuit might consist of a pair of 1.5mm² cables feeding a few lights or a very large three-core cable feeding a large motor direct from a circuit breaker or the main switchboard.

Regulations Governing Final Circuits

IEE Regulation 314.4 states that 'where an installation comprises more than one final circuit, each shall be connected to a separate way in a distribution board', and that the wiring to each final circuit shall be electrically separated from that of every other final circuit.

For final circuits the nominal current rating of the fuse or circuit breaker (overcurrent device) and cable will depend on the type of final circuit. Final circuits can be divided into the following types, all of which will need different treatments when planning the size of the conductors and the rating of the overcurrent devices:

- Final circuit feeding 13A sockets to BS 1363,
- Final circuit feeding sockets to BS EN 60309-2 (industrial types 16A to 125A),
- Final circuit feeding fluorescent or other types of discharge lighting,
- Final circuit feeding motors and
- Final circuit feeding cookers.

Final Circuit Feeding 13A Sockets to BS 1363

The main advantages of the 13A socket with fused plug are that any appliance with a loading not exceeding 3kW (13A at 230V unity Power Factor) may be

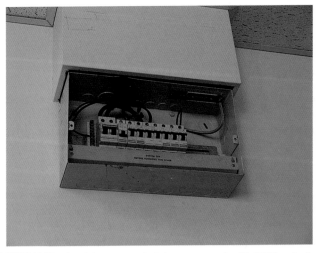

FIGURE 5.19 An eight-way metal-clad consumer unit with MCB protection.

FIGURE 5.20 Single and twin 13A socket outlets can be obtained in all-insulated or metal-clad forms, to allow appropriate equipment selection to suit site conditions.

connected with perfect safety to any 13A socket. Under certain conditions an unlimited number of sockets may be connected to any one circuit.

One point which must be borne in mind by the designer is the question of the use of outdoor equipment. IEE Regulation 411.3.3 states that where a socket outlet may be *expected* to supply portable equipment for use outdoors, it shall be protected by an RCD with a rated residual current not exceeding 30mA. RCDs are also an IEE requirement in several other circumstances and information on this is detailed in the Regulations.

Circuit arrangements Recommendations exist in Appendix 15 of the IEE Regulations for standard circuit arrangements with 13A sockets. These permit 13A sockets to be wired on final circuits as follows (subject to any de-rating factors for ambient temperature, grouping or voltage drop):

- A number of socket outlets connected to a final circuit serving a floor area not exceeding 100m² wired with 2.5mm² PVC insulated cables in the form of a ring and protected by a 30A or 32A overcurrent protective device.
- A number of socket outlets connected to a final circuit serving a floor area not exceeding 75m² with 4mm² PVC cables on a radial circuit and protected by an overcurrent device of 30A or 32A rating.
- A number of socket outlets connected to a final circuit serving a floor area not exceeding 50m² with 2.5mm² PVC cables on a radial circuit and protected by an overcurrent device not exceeding 20A.

Spurs may be connected to these circuits. If these standard circuits are used the designer is still responsible for ensuring that the circuit is suitable for the expected load. Also the voltage drop, and earth fault loop impedance values are suitable and the breaking capacity of the overload protection is sufficiently high.

If the estimated load for any given floor area exceeds that of the protective device given above then the number of circuits feeding this area must be increased accordingly.

Spurs

Non-fused spurs: A spur is a branch cable connected to a 13A circuit. The total number of non-fused spurs which may be connected to a 13A circuit must not exceed the total number of sockets connected directly to the circuit. Not more than one single or one twin socket outlet or one fixed appliance may be connected to any one spur. Non-fused spurs may be looped from the terminals of the nearest socket, or by means of a joint box in the circuit. The size of the cable feeding non-fused spurs must be the same size as the circuit cable.

Fused spurs: The cable forming a fused spur must be connected to the ring circuit by means of a 'fused connection unit'. The rating of the fuse in this unit shall not exceed the rating of the cable forming the spur, and must not exceed 13A.

There is no limit to the number of fused spurs that may be connected to a ring. The minimum size of cables forming a fused spur shall be 1.5mm^2 PVC with copper conductors, or 1.0mm^2 MI cables with copper conductors.

Fixed appliances permanently connected to 13A circuits (not connected through a plug and socket) must be protected by a fuse not exceeding 13A and a double pole (DP) switch or a fused connection unit which must be separated from the appliance and in an accessible position.

When planning circuits for 13A sockets it must always be remembered that these are mainly intended for general purpose use and that other equipment such as comprehensive heating installations, including floor warming, should be circuited according to the connected load, and should not use 13A sockets.

Fuselinks for 13A plugs: Special fuselinks have been designed for 13A plugs; these are to BS 1362 and are standardised at 3A and 13A, although other ratings are also available.

Flexible cords for fused plugs

for 3A fuse 0.50mm^2
for 13A fuse 1.25mm^2

All flexible cords attached to portable apparatus must be of the circular sheathed type, and not twin twisted or parallel type. With fused plugs, when a fault occurs resulting in a short circuit, or an overload, the local fuse in the plug will operate, and other socket outlets connected to the circuit will not be affected. It will be necessary to replace only the fuse in the plug after the fault has been traced and rectified.

13A Circuit for Non-Domestic Premises

For industrial, commercial and similar premises the same rules apply as for domestic premises in as much as the final circuit cables must be protected by suitable overcurrent devices.

It is often necessary, however, to connect a very large number of sockets to a single circuit, many more than would be recommended for domestic premises. For example, in a laboratory it may be necessary to fit these sockets on benches at frequent intervals for the sake of convenience. The total current required at any one time may be comparatively small and therefore a 20A radial or ring circuit, protected by a 20A fuse or circuit breaker, and wired with 2.5mm^2 PVC cables, could serve a large number of sockets. In this case the area being served must be in accordance with the standard circuit arrangements given in the IEE On-site guide.

Final Circuit for Socket Outlets to BS EN 60309

These socket outlets are of the heavy industrial type, and are suitable for single-phase or three-phase with a scraping earth. Fuses are not fitted in the sockets or the plugs. Current ratings range from 16A to 125A.

The 16A sockets, whether single- or three-phase, may be wired only on radial circuits. The number of sockets connected to a circuit is unlimited, but the protective overcurrent device must not exceed 20A. It is obvious that if these 16A sockets are likely to be fully loaded then only one should be connected to any one circuit. The higher ratings will of course each be wired on a separate circuit. Due to their robust nature these sockets are often used in industrial installations to feed small three-phase motors, and if the total estimated load of the motors does not exceed 20A then there is no reason why a considerable number should not be connected to one such circuit.

FIGURE 5.21   Industrial plug and socket BS EN 60309-2.

The same rule which applies to all final circuits must be complied with, which is that the conductors and protective devices must be suitably rated as already explained.

Final Circuits Feeding Fluorescent and Other Types of Discharge Lighting

Discharge lighting may be divided into two groups: those which operate in the 200V/250V range, and the HV type which may use voltages up to 5000V to earth. The first group includes tubular fluorescent lamps which are available in ratings from 8W to 125W, high- and low-pressure sodium lamps which are rated from 35W to 400W, also high- and low-pressure mercury vapour lamps rated from 80W to 1000W, and other forms of discharge lighting.

The second group includes neon signs and similar means of HV lighting.

LV discharge lighting circuits: Regulations governing the design of final circuits for this group are the same as those which apply to final circuits feeding tungsten lighting points, but there are additional factors to be taken into account. The current rating is based upon the 'total steady current' which includes the lamp, and any associated control gear, chokes or transformers, and also their harmonic currents. In the absence of manufacturers' data, this can be arrived at by multiplying the rated lamp power in Watts by 1.8, and is based on the assumption that the power factor is not less than 0.85 lagging. It should be noted that current fluorescent technology utilising High Frequency control gear and High Efficiency lamps run with a power factor close to unity, but the harmonic content of the supply will still need to be considered. Manufacturers generally publish data detailing the recommended CPDs that best serve their luminaires.

The control gear for tubular fluorescent lamps is usually enclosed in the casing of the luminaire, but for other types of discharge lighting, such as high-pressure mercury and sodium, the control gear is sometimes mounted remote from the luminaire. Here it is necessary to check the current which will flow between the control gear and the lamp. The remote control gear must be mounted in a metal box, must be provided with adequate means for the dissipation of heat, and spaced from any combustible materials.

Another disadvantage of locating control gear remote from discharge lamps is that, if a fault develops in the wiring between the inductor and the lamp, the presence of the inductor will limit the fault current so that it may not rise sufficiently to operate the fuse. Such a fault could very well remain undetected. If any faults develop in these circuits this possibility should be investigated.

Circuit switches: Circuit switches controlling fluorescent and discharge circuits should be designed for this purpose otherwise they should be rated at twice that of the design current in the circuit. Quick-break switches must not be

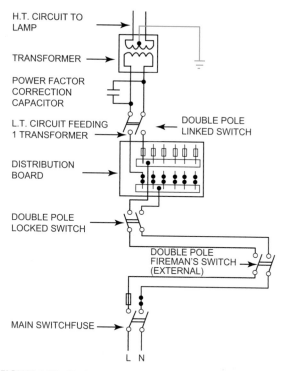

FIGURE 5.22 Typical circuit feeding h.t. electric discharge lamps.

used as they might break the circuit at the peak of its frequency wave, and cause a very high induced voltage which might flash over to earth.

Another way to overcome this issue is by switching the lighting via a contactor arrangement, which is controlled via a separate switching circuit. This proves useful when controlling large numbers of luminaires from single or multiple/remote locations, it also provides a great amount of flexibility as automatic and centralised control system can be employed if required.

Three-phase circuits for discharge lighting: In industrial and commercial installations it is sometimes an advantage to split the lighting points between the phases of the supply, and to wire alternate lighting fittings on a different phase. This enables balancing of the load and ensures that the loss of a single phase allows reduced lighting level over the whole area with the remaining phases operating. When wiring such circuits it is preferable to provide a separate neutral conductor for each phase, and not wire these on three-phase four-wire circuits. The reason for this is that for this type of lighting very heavy currents may flow in the neutral conductors, due to harmonics and/or imbalances between phases. Luminaires connected on different phases must be provided with a warning notice DANGER 400V on each luminaire.

Stroboscopic effect: This is not a problem with the high frequency lighting which is generally available but in the past one disadvantage of discharge lighting was the stroboscopic effect of the lamps. This was caused by the fact that the discharge arc was actually extinguished 100 times per second with a 50Hz supply. There was a danger in that it could make moving objects appear to be standing still, or moving slowly backwards or forwards when viewed under this type of lighting.

HV Discharge Lighting Circuits

HV is defined as a voltage in excess of LV, i.e. over 1000V a.c. The IEE Regulations generally cover voltage ranges only up to 1000V a.c., but Regulation 110.1 also includes voltages exceeding LV for equipment such as discharge lighting and electrostatic precipitators.

Discharge lighting at HV consists mainly of neon signs, and there are special regulations for such circuits. The installation of this type of equipment is usually carried out by specialists. The equipment must be installed in accordance with the requirements of British Standard BS 559, 'Specification for design, construction and installation of signs'.

Final Circuits Feeding Cookers

In considering the design of final circuits feeding a cooker, diversity may be allowed. In the household or domestic situation, the full load current is unlikely to be demanded. If a household cooker has a total loading of 8kW the total current at 230V will be 34.8A, but when applying the diversity factors the rating of this circuit will be:

first 10A of the total rated current	= 10.0A
30% of the remainder	= 7.4A
5A for socket	= 5.0A
Total	= 22.4A

Therefore the circuit cables need only be rated for 22.4A and the overcurrent device of similar rating.

Cookers must be controlled by a switch which must be independent of the cooker. In domestic installations this should preferably be a cooker control unit which must be located within 2m of the cooker and at the side so that the control switch can be more easily and safely operated.

Pilot lamps within the cooker control unit need not be separately fused. Reliance must not be placed upon pilot lamps as an indication that the equipment is safe to handle.

FIGURE 5.23 Old and new style cooker control units, incorporating a 13A socket outlet. These units are also available with neon indicator lights.

5.7 CIRCUITS SUPPLYING MOTORS

Final Circuits Feeding Motors

Final circuits feeding motors need special consideration, although in many respects they are governed by the regulations which apply to other types of final circuits. The current ratings of cables in a circuit feeding a motor must be based upon the full load current of the motor, although the effect of starting current will need to be considered if frequent starting is anticipated [IEE Regulation 552.1.1]. Every electric motor exceeding 0.37kW shall be provided with control equipment incorporating protection against overload of the motor. Several motors not exceeding 0.37kW each can be supplied by one circuit, providing protection is provided at each motor.

Motor Isolators

All isolators must be 'suitably placed' which means they must be near the starter, but if the motor is remote and out of sight of the starter then an additional isolator must be provided near the motor. All isolators, of whatever kind, should be labelled to indicate which motor they control.

The cutting off of voltage does not include the neutral in systems where the neutral is connected to earth. For the purposes of mechanical maintenance, isolators enable the person carrying out maintenance to ensure that all voltage is cut off from the machine and the control gear being worked upon, and to be certain that it is not possible for someone else to switch it on again inadvertently. Where isolators are located remote from the machine, they should have removable or lockable handles to prevent this occurrence.

Motor Starters

It is necessary that each motor be provided with a means of starting and stopping, and so placed as to be easily worked by the person in charge of the motor. The starter controlling every motor must incorporate means of ensuring that in the event of a drop in voltage or failure of the supply, the motor does not start automatically on the restoration of the supply, where unexpected re-starting could cause danger. Starters usually are fitted with undervoltage trips, which have to be manually reset after having tripped.

Every motor having a rating exceeding 0.37kW must also be controlled by a starter which incorporates an overcurrent device with a suitable time lag to look after starting current [IEE Regulation 552.1.2]. These starters are generally fitted with thermal overloads which have an inherent time lag, or with the magnetic type which usually have oil dashpot time lags. These time lags can usually be adjusted, and are normally set to operate at 10% above full load current. Electronic protective relays are also available and these provide a fine degree of protection.

Rating of protective device IEE Regulation 433.2.2 states that the overcurrent protective device may be placed along the run of the conductors (provided no branch circuits are installed), therefore the overcurrent protective device could be the one incorporated in the starter, and need not be duplicated at the commencement of the circuit.

Short-circuit protection must be provided to protect the circuit, and shall be placed where a reduction occurs in the value of the current-carrying capacity of the conductors of the installation (i.e. such as in a distribution board). The device may, however, be placed on the load side of a circuit providing the conductors between the point where the value of the current-carrying capacity is reduced and the position of the protective device does not exceed 3m in length and providing the risk of fault current, fire and danger to persons is reduced to a minimum [IEE Regulation 433.2.2].

When motors take very heavy and prolonged starting currents it may well be that fuses will not be sufficient to handle the starting current of the motor, and it may be necessary to install an overcurrent device with the necessary time delay characteristics, or to install larger cables.

With three-phase motors, if the fuses protecting the circuit are not large enough to carry the starting current for a sufficient time, it is possible that one may operate, thus causing the motor to run on two phases. This could cause serious damage to the motor, although most motor starters have inherent safeguards against this occurrence.

The ideal arrangement is to back up the overcurrent device in the motor starter with HRC fuselinks which have discriminating characteristics which will carry heavy starting currents for longer periods than the overload device. If there is a short circuit the HRC fuses will operate and clear the short circuit before the short circuit kVA reaches dangerous proportions.

Slip-ring motors: The wiring between a slip-ring motor starter and the rotor of the slip-ring motor must be suitable for the starting and load conditions. Rotor circuits are not connected directly to the supply, the current flowing in them being induced from the stator. The rotor current could be considerably greater than that in the stator; the relative value of the currents depending upon the transformation ratio of the two sets of windings.

The cables in the rotor circuit must be suitable not only for full load currents but also for starting currents. The reason is that, although heavy starting currents may only be of short duration (which the cables would easily be able to carry), if the cables are not of sufficient size to avoid a voltage drop this could adversely affect the starting torque of the motor.

The resistance of a rotor winding may be very low, and the resistance in the rotor starter is carefully graded so as to obtain maximum starting torque consistent with a reasonable starting current. If cables connected between the rotor starter and the rotor are fairly long and restricted in size, the additional resistance of these cables might even prevent the motor from starting. When slip-ring motors are not fitted with a slip-ring short-circuiting device, undersized rotor cables could cause the motor to run below its normal speed.

Before wiring rotor circuits always check the actual rotor currents, and see that the cables are of sufficient size so as not to adversely affect the performance of the motor.

Emergency Switching

IEE Regulation 537.4.1.1 states that 'means shall be provided for emergency switching of any part of an installation where it may be necessary to control the supply to remove an unexpected danger'.

Generally it is desirable to stop the motor which drives the machine, and if the 'means at hand' is not near the operator then STOP buttons should be provided at suitable positions (Fig. 5.24), and one must be located near the operator, or operators. Stop buttons should be of the lock-off type so that the motor cannot be restarted by somebody else until such time as the stop button which has been operated is deliberately reset.

In factory installations it is usual to provide stop buttons at vantage points throughout the building to enable groups of motors to be stopped in case of emergency. These buttons are generally connected so as to control a contactor which controls a distribution board, or motor control panels. For a.c. supplies stop buttons are arranged to open the coil circuit of a contactor or starter. For d.c. supplies the stop buttons are wired to short circuit the hold-on coil of the d.c. starter.

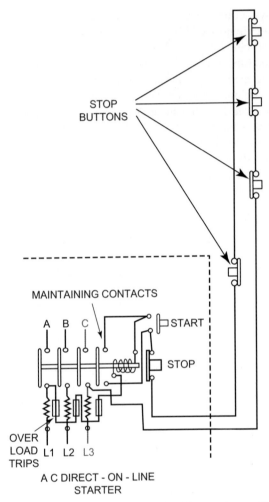

STOP
BUTTONS

MAINTAINING CONTACTS

A   B   C

START

STOP

OVER
LOAD   L1   L2   L3
TRIPS

A C DIRECT - ON - LINE
STARTER

FIGURE 5.24   Safety precaution. Means must be at hand for stopping machines driven by an electric motor. One method of doing this is to fit remote STOP buttons at convenient positions.

Reversing Three-Phase Motors

When three-phase motors are connected up for the first time it is not always possible to know in which direction they will run. They must be tested for direction of rotation. If the motor is connected to a machine, do not start it if there is a possibility that the machine may be damaged if run in the wrong direction. If the motors run in the wrong direction it is necessary only to change over any two wires which feed the starter (L1, L2 and L3).

In the case of a star delta starter, on no account change over any wires which connect between the starter and the motor because it is possible to change over the wrong wires and cause one phase to oppose the others.

FIGURE 5.25 Emergency stop buttons, here shown with and without a key operated reset facility. The former (shown left) can be used where restoring the power must be carried out by authorised persons.

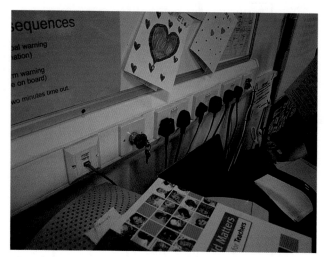

FIGURE 5.26 Another situation where emergency buttons may be required is in locations where young people may be present. The facility is provided in this school classroom and current is disconnected a contactor which is operated by the emergency button. A key reset facility is provided.

For slip-ring motors it is necessary only to change over any two lines feeding the starter, it is not necessary to alter the cables connected to the rotor. To reverse the direction of single-phase motors it is generally necessary to change over the connections of the starting winding in the terminal box of the motor.

Lifts

Electrical installations in connection with lift motors must comply with BS EN 81-2.

The actual wiring between the lift control gear and lift is carried out by specialists, but the designer of the sub-main needs to comply with the requirements of BS EN 81-2. The power supply to a lift or to a lift room, which

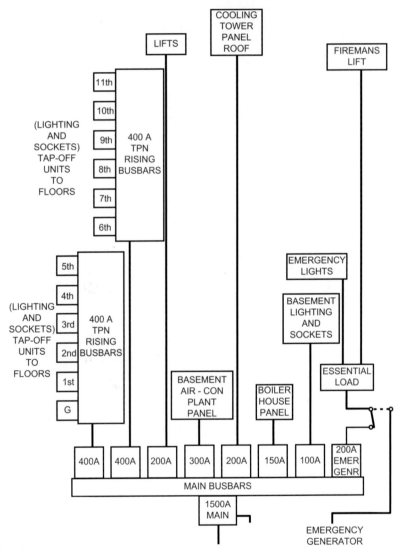

FIGURE 5.27 Distribution diagram for typical commercial multi-storey building – meters on each floor if required.

may control a bank of lifts, must be fed by a separate distribution cable from the main switchboard. The cable must be of such a size that for a three-phase 400V supply the voltage drop must not exceed 10V when carrying the *starting* current of the lift motor or the motor generator. This is the usual maximum volt drop specified by lift manufacturers.

The main switchgear should be labelled LIFTS and in the lift room circuit breakers or a distribution board must be provided as required by the lift manufacturers.

The supply for the lift car light must be on a separate circuit. It is usual to provide a local distribution board in the lift motor room and the lights controlled by a switch in the lift motor room. These cables must be entirely separated from the cables feeding the power supply to the lift. These lights should be connected to a maintained/emergency supply, so that in the event of mains failure the lights in the lift cage are not affected. Alarm systems should also be connected to a maintained/emergency supply or from a battery.

Cables other than those connected to lift circuits must not be installed in lift shafts, but cables connected to lift circuits need not necessarily be installed in lift shafts.

Where determined by the fire engineering report/local fire officer, certain buildings may require a designated fireman's lift, this could be on a separate fire protected circuit with an change-over arrangement to restore the supply via an essential generator back supply in the event of a main failure, so that in the event of a fire the supply to this lift is maintained when other supplies are switched off. The secondary supply may need to be rerouted diversely from the primary supply, and the requirements for this supply are very specific and may need to comply with BS 5588.

Chapter 6

Worked Example

Electrical design processes and principles have been covered in the preceding chapters and, to give an idea as to the application of the techniques, the following worked example is offered. In the example, a client is to build a commercial unit as part of a development. The performance of the Mechanical Electrical & Public Health (MEP) services has already been defined, and the example is provided to examine the design of the electrical building services. Details of the building are shown in Fig. 6.1.

The unit is to comprise a warehouse complete with an office area with provisions of public health services, such as WC etc. In the example, the developer will also provide the external services for a car park and delivery area. The outline floor areas are as defined in Table 6.1.

In reality, the outline scope of works would be to design, install and test the complete electrical services, meeting the appropriate standards and regulations. The scheme would need to be approved by the client, and then be submitted to statutory authorities for approval. The documentation produced to reach this stage of development by the contractor would run to many pages, but in this example only the main electrical issues are addressed.

6.1 DESIGN CRITERIA

Typical criteria for an installation of this type will be used in the example and these are defined in Table 6.2.

In the example, a set of room data sheets has also been provided to outline the individual employer requirements, and these are summarised in Table 6.3. Further details of the lighting types proposed can be found in Table 6.4.

Particular Specification

In any required installation, the particular specification of the services may well be defined, some examples being:

- 20% Spare ways and capacity – blanking plates, segregated lighting and power Distribution Boards;
- Sub-main cabling to be Cu PVC/XLPE/SWA/LSF fixed on medium duty cable tray/ladder;

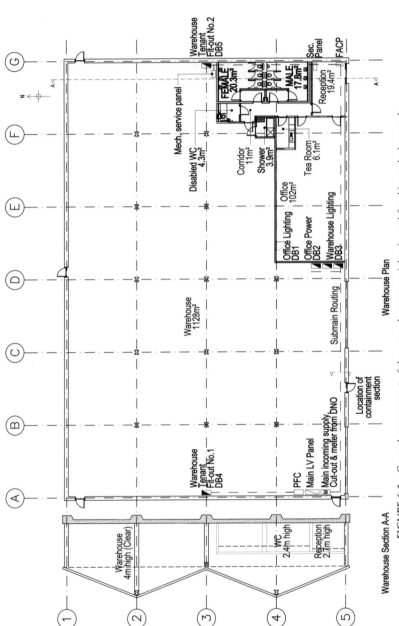

FIGURE 6.1 General arrangement of the example commercial unit used for this worked example.

TABLE 6.1 Outline Floor Areas

Area	Measured area from drawings (m^2)
Reception	19.4
Office	102
Tea room	6.1
Corridor	11
Male WC	17.8
Female WC	20.3
Disabled WC	4.3
Shower	3.9
Warehouse	1128
Total	1312.8

- Check meters to comply with Building Regulations Part L2;
- Distribution Boards to be wall mounted, metal clad and hinged with lockable covers;
- Electrical Services that are required in relation to mechanical services, e.g. supplies to boilers, control panels and so on;
- Electronic systems including CCTV and security to be provided;
- Emergency lighting to be provided;
- Earthing and bonding in line with BS7671;
- Fire alarms to be provided in line with BS 5839;
- Lighting to utilise high-efficiency luminaries;
- Lighting protection is to be provided complete with Electronic Surge Protection (ESP) units;
- MCCB main panel boards are to be wall mounted Form 4 type 2 panels;
- Future fit-out allowances to be made for the tenant including spare ways within distribution boards;
- Small power to be provided as detailed within the design criteria; and
- The power factor is to be corrected to 0.95.

6.2 PROCESS OF DESIGN

The design process has been fully described in the previous chapters of this book and the results which relate to this worked example are covered in the paragraphs below.

TABLE 6.2 Design Criteria

Criteria	Value
Voltage	400 V/230V
Phases	Three phase and neutral four wire
Frequency	50Hz
Overall power factor	0.8
Earthing system	PME
External earth loop impedance (Z_e)	To be defined by the REC
Prospective Short-Circuit Current (PSCC)	To be defined by the REC
Calculation reference	BS7671:2008
Small power loading[a]	25W/m^2
Lighting loading[a]	15W/m^2
Tenant fit-out load allowance	90W/m^2
Mechanical plant loading[b]	40W/m^2
Future capacity required	20%
Ambient temperature	30 °C
External temperatures	29 °C Summer
	−3 °C Winter

[a]Based on CIBSE guide F Section 12.2.2 (Small Power), 9.6 (lighting).
[b]Based on CIBSE guide K Section 3.3.3. – figure doesn't include warehouse (covered in fit-out allowance).

To determine the location and sizing of switchgear and to carry out preliminary sizing and routing of cables, an initial assessment of maximum demand and calculations of the approximate main cable sizes are required.

Assessment of Maximum Demand

Referring to the maximum demand calculation in Chapter 4 of this book and the previous information gained, the initial maximum demand can be estimated as shown in Table 6.5.

From the above, it can be seen that the assessed total connected load is approximately 133kW. At this point the overall diversity and an allowance for

TABLE 6.3 Summarised Room Data Sheets

Room	Lighting level (Lux)	Lighting level measured at	Uniformity	Luminaire type (see Table 6.4)	Finishes	Occupancy	Special requirements
Office	350	0.8m[a]	80%	A	White PVC	1 Person/10m^2	DSSO per person arranged on Dado trunking at 3m intervals per 2 No. DSSO
Entrance	200	FFL	80%	B	Brushed steel	2 Persons	3 No. DSSO Floor Box for reception desk inc 2 No. DSSO & RJ45 Powered Entrance door
Male and Female WCs	100	FFL	80%	C and D	White PVC	N/A	Hand drier/shaver outlet
Disabled WC/shower room	100	FFL	80%	F	White PVC IP 54	N/A	Supply to shower disabled call alarm
Tea area	300	Worktop	80%	A and G	White PVC	N/A	Power for microwave/fridge kettle etc.
Circulation Areas	100	FFL	80%	B	White PVC	N/A	Power for cleaners
Tenant fit-out	300	FFL	80%	H	Metalclad	N/A	Power to doors etc. All other power by tenant
Plant Area	150	FFL	80 %	I	Metalclad	N/A	

DSSO = Double switched socket outlet.
[a]Measurement taken from Finished Floor Level (FFL).

TABLE 6.4 Lighting Types

Luminaire Type	Type	Mounting	Control gear	Control	Notes
A	Modular fluorescent	Recessed	HF and Dimmable[a]	PIR and Daylight linked	Direct/indirect distribution to meet CIBSE guide LG 7
B	Circular compact fluorescent	Recessed	HF	Switched	Decorative attachment required
C	Circular compact fluorescent	Recessed	HF	PIR and Timer	
D	LED down light	Recessed	N/A	PIR and Timer	To be located over basins
E	Reserved for Emergency lighting references				
F	Circular compact fluorescent	Recessed	HF	PIR and Timer	To be protected to IP54, zones to be observed
G	Circular compact fluorescent	Recessed	HF	Switched	
H	Low-bay discharge	Surface	HF	Switched	To be mounted at a minimum of 4m AFFL
I	Linear fluorescent	Surface	HF	Switched	IP65 Anti-corrosive

HF = High frequency.
[a]Luminaries to be daylight linked to allow output to be reduced if sufficient daylight is available.

future capacity (as stated within the design criteria) may be considered to arrive at the declared maximum demand. This is shown in Table 6.6.

The overall Power Factor (PF) is calculated on 0.8 in accordance with the design criteria, but 0.95 has been used to calculate the maximum demand. Power Factor Correction (PFC) equipment will be installed and this will correct the figure imposed on the District Network Operators (DNO) supply. (Further details are given in the PFC section in Chapter 5.)

Note that if the 128kW supply was not corrected to 0.95 PF lagging, then the declared maximum demand would be 160kVA, as:

$$P = \sqrt{3}\ VI \cos \theta \quad \text{therefore } VI = P/\cos \theta$$

Where $\cos \theta = 0.8$, the kVA is $128\text{kW}/0.8 = 160\text{kVA}$ (230.95A/phase).

Where $\cos \theta = 0.95$, the kVA is $128\text{kW}/0.95$

$\quad = 134.7\text{kVA}$ (194.43A/phase).

The declared maximum demand is specified to the electricity supplier in kVA as this figure is independent of the fluctuations in power factor or nominal voltage, and allows a fixed amount of current to be supplied at a nominal

TABLE 6.5 Assessment of Maximum Demand

Area by type	Total area (m²)	Typical loading (W/m²)	Total loading (W)
Reception/office	138.5	40[a]	5540
WCs	46.3	25[b]	1158
Warehouse	1128.0	105[c]	118,440
Mechanical Services	184.8[d]	40[a]	7392
		Total	132,530

[a]$40\text{W/m}^2 = 15\text{W/m}^2$ for lighting and 25W/m^2 for power (based on design criteria).
[b]$25\text{W/m}^2 = 15\text{W/m}^2$ lighting (based on design criteria) and an allowed 10W/m^2 power (assumed the power requirement is not as great).
[c]$105\text{W/m}^2 = 90\text{W/m}^2$ for the fit-out loading and 15W/m^2 for the lighting (based on design criteria) the general small power is included within the fit-out figure.
[d]Area is sum of Reception/office and WCs

TABLE 6.6 Declared Maximum Demand

Total Connected load	133kW
Applicable overall diversity	0.8PU (80%)
Future Capacity[a]	1.2PU (20%)
Total expected maximum demand	128kW
Corrected Power Factor[b]	0.95
Declared maximum demand[c]	135kVA

[a]Future capacity as specified design criteria.
[b]The corrected Power Factor is used as defined in the specific project requirements.
[c]This figure is commonly specified in kVA.

voltage. It also allows the power required to be transposed as a common unit between single, three phase or higher nominal voltage supplies.

In this example, which requires a 135kVA supply, the likely supply provided by the DNO would be 150kVA (as this is the next 'standard' higher supply size).

Locating the Incoming Point of Supply (POS)

The DNO can provide the requirements that need to be met when determining the location of their equipment, and co-ordination between the DNO and the professional parties involved must take place when determining the incoming POS location. The general principles for locating the intake position are described in Chapter 5. In this example, this equipment is to be situated in the south-west corner of the warehouse.

Location of Principle Distribution Equipment and Switchgear

The next step is to determine where the remainder of the distribution equipment is to be located. First, it is necessary to establish what distribution equipment is required.

The building can be split into logical areas. These could be determined by physical constraints or by grouping areas of similar usage. The type of electrical loads can play a part, and compliance with Building Regulations Part L may require the distribution to be split into lighting, office power and mechanical services so as to assist with the sub-metering. There may also be financial considerations as to how far the distribution is split for metering purposes.

Figure 6.1 shows the building which includes an office, welfare facilities and warehouse. With a speculative development, a basic level of electrical installation should be provided to the reception, office and WC plus statutory services such as emergency warehouse lighting and the fire alarm system. Typically, the tenant will provide all other services.

A suggested split of distribution could be:

- Office and amenity area lighting
- Office and amenity area power (including roller shutter door within the warehouse and external power supplies forming part of the base build)
- Warehouse lighting (including external lighting)
- Spare ways for tenant fit-out equipment

Other supplies requiring consideration could typically be:

- Mechanical building services
- Fire detection and alarm system
- Security systems
- Electronic Surge Protection (ESP) equipment
- Power Factor Correction (PFC) equipment

The location of these additional items will generally be determined by the location of their control panels or equipment. An exception would be the ESP and PFC equipments which need to be located adjacent to the main switch panel. Therefore the main items of final distribution can be located as follows.

Office lighting and office power distribution boards: could be located external to the office (to allow easy maintenance access without hindering the office layout).

Warehouse lighting distribution board: could be central to the warehouse to minimise the length of cable runs plus to optimise the length of the sub-main cabling.

Tenant fit-out distribution board: could be located at the centre of the warehouse to provide the most balanced lengths of final circuit cabling.

Fire detection and alarm panel: to be located adjacent to the main entrance where the fire service will enter as recommended by BS5839-1 (23.2.1) and therefore located within the reception area.

Ventilation and mechanical equipment: to be located adjacent to the amenities block where the load will be concentrated.

Security control panel: could be located within the reception adjacent to the point of entry.

Power factor correction equipment: would be located adjacent to the Main switch panel as this is the point of utilisation.

At this stage, an assessment of the approximate loading should be made to ensure that the suggested distribution areas are practicable. The areas detailed in Table 6.1 can be combined with the W/m^2 values from Table 6.5 to give the result shown in Table 6.7.

From an examination of the detail above, all the supplies can be fed from a single Distribution Board (DB) with the exception of the warehouse fit-out DB which would require its own 220 A supply. This would not co-ordinate with the upstream Circuit Protective Device (CPD). A solution would be to split the fit-out DB into two boards to reduce the prospective load on each. In this case it would also be appropriate to locate the distribution boards at either end of the warehouse to reduce the concentrated loads and the length of the final circuit cabling. The locations of these items are shown in Fig. 6.1. The actual loading and number of ways for each distribution board would be determined at a later stage.

Determining the Principle Containment Routes

The next step is to select the principle containment routes for the distribution cabling. These routes can be influenced by a number of criteria, but in general it is preferential that the shortest routes be followed, thus reducing the cable lengths. This, in turn, will reduce the voltage drop (VD), loop impedances and therefore cable sizes. However, the routes need to be co-ordinated with the

TABLE 6.7 Approximate Loadings

	m²	W/m²	Total (W)	Including future capacity[a] (kW)
Office lighting distribution board Reception/office etc.	138.5	25	3462.5	
WCs	46.3	10	463.0	
		Total	3925.5	4.7
Office power distribution board Reception/office etc.	138.5	15	2077.5	
WCs	46.3	15	694.5	
		Total	2772.0	3.3
Office power distribution board Warehouse lighting distribution board	1128	15	16920	20.3
Office power distribution board Warehouse fit-out distribution board	1128	90	101520	121.8

[a]20% Additional capacity included as stated within the design criteria.

building structure, any clear heights that require to be maintained, segregation from other services [IEE Regulation 528] and voltage bands.

Once the routing of the containment has been determined, the lengths of the sub-main distribution cabling can be measured from the drawing. The measurements need to consider any changes in level, drops and risers to and from equipment, and cabling required for the termination. The lengths measured for our example are shown in Table 6.8.

Estimation of Distribution and Switchgear Equipment Loadings

Now that the distribution principles have been determined, this information can be combined with the assessment of maximum demand figure previously attained to give an estimation of the expected connected loads on each of the sub-mains. From this information the anticipated load on each Distribution Board and sub-circuit can be compiled as detailed in Table 6.9.

At this point it would be usual to decide which loads are to be supplied as single phase and which are to be three phase, each circuit being considered in turn.

Office lighting and power DBs have very little total connected load. A three-phase DB is sometimes used to balance the loads and keep the sub-main cable sizes down, but in this case a TPN supply is not warranted, so the use of an SPN supply would be possible. This will also assist in minimising the potential for three phases being present in the lighting switches and across the PCs in the office area.

TABLE 6.8 Sub-Main Cable Lengths

Item Number	Supplying	Location	Length (m)
1	Office Lighting Distribution Board	External to Office	39
2	Office Power Distribution Board	External to Office	39
3	Warehouse Lighting Distribution Board	Central to the warehouse	40
4	Warehouse Tenant Fit-out Distribution Board No. 1	West end of warehouse	21
5	Warehouse Tenant Fit-out Distribution Board No. 2	East end of warehouse	71
6	Fire detection and alarm control panel	In reception area	55
7	Ventilation / Mechanical Services control panel	Adjacent amenity block	68
8	Security Control Panel	At main entrance	59
9	Power Factor Correction (PFC) equipment	Adjacent main panel	2
10	Electronic Surge Protection (ESP) equipment	At main panel	0.5

Warehouse lighting DB could also be possibly fed from an SPN supply, but this would impose an imbalance between the phases of the main supply. The use of a TPN supply will allow the circuits to be balanced out more evenly, therefore a TPN DB should be utilised.

Warehouse power DB necessitates a large supply and this warrants a TPN feed. This will also allow the tenant to utilise three-phase supplies in their equipment.

Fire Alarm Control Panel (FACP) and security panel supplies will require an SPN supply, as this is usually recommended by the manufacturer of the equipment. The minimal load may warrant a small size CPD, but it is necessary to ensure discrimination between the manufacturer fitted CPD, the fused connection unit CPD and the CPD supplying the circuit. The feed may be *via* a standard 10A radial circuit, although a fused spur could be used, fused at, say, 3A.

Another issue to be considered is that a 10A CPD may not be available as some manufacturers do not produce the smaller sized CPDs for main switch

TABLE 6.9 Anticipated Loads

Item Number	Supplying	Connected load (kW)	TPN Load/ phase (A)	SPN Load (A)
1	Office Lighting Distribution Board	4.7	8.5	25.5
2	Office Power Distribution Board	3.3	6.0	17.9
3	Warehouse Lighting Distribution Board	20.3	36.6	110.3
4	Warehouse Tenant Fit-out Distribution Board No. 1	60.9[a]	109.9	331.0
5	Warehouse Tenant Fit-out Distribution Board No. 2	60.9[a]	109.9	331.0
6	Fire detection and alarm control panel (FACP)	0.5[b]	0.9	2.7
7	Ventilation/mechanical services control panel	8.9[c]	16.1	48.4
8	Security Control Panel	0.5[b]	0.9	2.7
9	Power Factor Correction (PFC) equipment	N/A[b]		
10	Electronic Surge Protection (ESP) equipment	N/A[b]		
	Total		288.8 (if all TPN)	869.5 (if all SPN)

Note that the loadings of Distribution Boards do not account for the diversified figures but are the expected total connected load.
[a]Is the figure taken from Table 6.7 divided over the 2 No. Distribution Boards.
[b]Nominal Load imposed.
[c]Figure is taken from table 6.5 + 20% Future capacity. Power Factor of 0.8 used as design criteria.

panels. In addition, the Prospective Short-Circuit Current (PSCC) may be too great for the withstand currents that the smaller CPDs would require (this will need to be checked).

Ventilation and mechanical service control panels would normally warrant an SPN supply, but if there is a requirement to supply motors, a TPN supply may be required. Co-ordination with the mechanical services engineer would be needed.

To simplify this example the Power Factor Correction (PFC) and Electronic Surge Protection (ESP) requirements will be omitted, but both will require a TPN supply/connection.

6.3 SELECTION OF SWITCHGEAR

When selecting the switchgear there are a number of items, such as the manufacturer, the type, the rating, the CPDs to be used, the IP rating and form of the switchgear, that need to be considered. For ease of maintenance it would be usual practice to select all the switchgears from the same manufacturer.

Main Panel Board

At this stage of the design process, it is possible to determine the requirements for the main panel board which are four SPN supplies and six TPN supplies. This equates to 22SPN ways or an eight-Way TPN main panel. In addition, it is necessary to make an allowance for a future capacity of 20% as stated in the specific requirements as listed earlier, which would equate to 10TPN ways. From an examination of manufacturer's data, it is possible to determine the nearest larger TPN Panel Board available and this would be 12-Way TPN Panel Board. Note that as 20% is the minimum future capacity to be allowed, it may be preferable to add more spare ways at little additional cost, but give more spare capacity. It would be normal to leave spares as full TPN ways, allowing maximum flexibility for tenants at a later date.

Fitting out the main panel board would give the following arrangement detailed in Table 6.10. Note that the circuits have been rearranged to make the maximum number of spare ways available.

At this stage, the initial schematic diagram can be produced. This is shown in Fig. 6.2. The lengths can then be added to the schematic diagram. The actual lengths are to be confirmed by site measure later on in the process.

The results shown in Table 6.10 raise a few questions. Firstly the total connected loads per phase are not in balance. This is often the case, although as the loads are within 10% of each other this would be acceptable. Secondly the total connected loads are much greater than the estimates made. This is due to the overall diversity not being taken into account and the figures being based on a power factor of 0.8 for the installation, rather than the corrected PF applied to the supply.

From the assessment of anticipated maximum demand, and the details provided by the DNO, a 200A incoming supply would be required; therefore the main panel would also need to be rated at a minimum of 200A. However, to provide the maximum connected loads stated within Table 6.10 a minimum rated busbar of 300A would be required, which would relate to a 400A panel board, this being the manufacturer's next standard size.

For the purposes of this example an MCCB panel board has been selected from a preferred manufacturer's range which will provide the basic specification of the equipment including the fault rating, form, access and dimensions.

TABLE 6.10 Circuits Distributed over Main Panel

Circuit number	Phase	Supplying	Connected load (A)		
			L1	L2	L3
1	L1	Office lighting distribution board	25.5		
	L2	Office power distribution board		17.9	
	L3	Fire detection and alarm control panel (FACP)			2.7
2	L1	Warehouse lighting distribution board	36.6		
	L2			36.6	
	L3				36.6
3	L1	Warehouse Tenant Fit-out Distribution Board No. 1	109.9		
	L2			109.9	
	L3				109.9
4	L1	Warehouse Tenant Fit-out Distribution Board No. 2	109.9		
	L2			109.9	
	L3				109.9
5	L1	Ventilation/mechanical services control panel	16.1		
	L2			16.1	
	L3				16.1
6	L1	Security control panel	2.7		
	L2	Spare			
	L3	Spare			
		Total connected load (A)	300.7	290.4	275.2

Note: Ways 7–12 not shown (including supplies to PFC and ESP).

Final Distribution Boards

For the final distribution boards, MCBs have been selected for the outgoing ways for a number of reasons. They provide greater user flexibility, being easy to reset, RCDs and other accessories are readily available and they provide simpler discrimination with the upstream MCCBs of the main panel board.

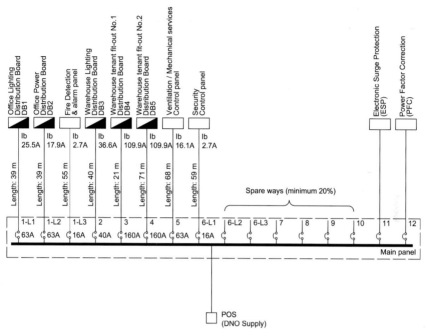

FIGURE 6.2 Schematic diagram of the sub-main distribution circuits.

Following on from this, the ratings and number of phases for each of the other distribution boards can be determined, and these are summarised in Table 6.11.

It is suggested that a selection of either 12SPN or 12TPN Way DB is made initially, which can be confirmed when the number of outgoing ways from each DB is finally decided.

A 250A distribution board has been selected for the warehouse so as to give maximum flexibility for the tenant. The actual connected peak load may exceed the 125A rating of a standard distribution board depending on what equipment the tenant may fit and a 250A rated board will usually give more space for terminating larger cables and outgoing supplies. Again the manufacturer can provide details of the basic specification of the equipment including the fault rating, form, access, maximum cable capacity and dimensions.

Spatial Requirements for Switchgear

Now that the initial selection of the main distribution equipment has been made, it is possible to determine the spatial requirements for the switchgear.

Laying Out the DNO Equipment

Guidance will normally be provided by the DNO stating the amount of space required for their equipment, typically '600mm from finished floor level, with

TABLE 6.11 Initial Schedule of Final Distribution Boards

Ref	Distribution board	Connected load/phase (A)	Type	Rating (A)
DB1	Office lighting	25.5	SPN	125
DB2	Office power	17.9	SPN	125
DB3	Warehouse lighting	36.6	TPN	125
DB4	Warehouse Tenant Fit-out No. 1	109.9	TPN	250[a]
DB5	Warehouse Tenant Fit-out No. 2	109.9	TPN	250[a]

[a]This is the standard rating of the busbars from the manufacturer's ranges, which can be supplied by a lower rated CPD, thus downrating the supply.

a 1050mm high × 500mm wide × 750mm deep zone required for the equipment.' They may also provide a diagram and other information to ensure that their requirements are met. A typical metering cutout, as provided by the DNO, is illustrated in Fig. 6.2.

Main Panel Board and Final Distribution Boards

For the distribution equipment, the details and sizes gained from the manufacturer's data can be utilised to determine the spatial requirements for the equipment. The manufacturers may also provide advice as to specific requirements for access. Sufficient allowance should be made for the plant room space in accordance with the criteria set out in Chapter 2.

6.4 PRELIMINARY SUB-MAIN CABLE SIZING

For the purposes of this worked example, one sub-main circuit will be examined in detail. The same process will, of course, need to be applied to all the circuits in any particular design scheme.

As described in Chapter 4, the process for sizing cables is to:

1. Determine the design load,
2. Select a suitable sized CPD,
3. Apply any rating factors required to the CPD size to determine the required cable rating (so that the cable is rated to the maximum load available from the CPD including rating factors) and
4. Determine the required cable rating, selecting the cable based on the tabulated ratings in the Appendix of BS7671 for the correct installation method.

Once the cable size is selected, it is then necessary to ensure that the cable conforms to the required volt drop limits and its CPD will disconnect within the required time to afford sufficient protection based on the impedances with the circuit etc.

Like most things, a cable has a beginning, a middle and an end, and although source details are not considered until later on in the calculation process, it is always good to start at the beginning. In this case, as details will be examined for the sub-main cable originating from the POS, it is necessary to take a look at the Supply Characteristics.

Supply Characteristics

As mentioned previously, the DNO will have provided information on the supply characteristics. The DNO will also provide information on the requirements for the CPDs, earthing arrangements and service cable sizes (customer tails).

A typical extract is shown in Table 6.12.

In this example, the 150 kVA supply has the following requirements:

- Service cable size $95mm^2$ Al
- Minimum customer tail cable size $70mm^2$ Cu
- Minimum earthing conductor $35mm^2$ Cu
- Minimum equipotential bonding conductor $25mm^2$ Cu

Applying the detail to the worked example, this would give the characteristics of available supply (IEE Regulation 132.2) required as:

- Nominal voltage (to earth)   400V/230V
- The nature of current and frequency   AC 50Hz
- External earth fault loop impedance (Ze)   0.20 ohms
- Prospective short-circuit current (PSCC)   16kA
- The overcurrent protective device   200A BS88 Fuse
- Maximum demand (declared supply   150kVA
  capacity)
- System of supply   TN-C-S (PME)

This information can be used to form the basis of the detailed design calculations.

Determining the Design Current

Now that the supply characteristics are known, the next step is to look at the other end of the cable to determine the design load of the equipment to be supplied. In this case, the approximate maximum design currents for the circuits are already known, as shown previously in Table 6.10. At this stage the value of the actual connected loads is not known, as detail of the outgoing design of all the distribution boards and equipment is yet to be determined. For the moment, therefore, the design current will be based on the approximate maximum demand currents.

TABLE 6.12 Typical DNO Supply Details

Maximum agreed supply capacity	Cut-out size	Fuse size	Service cable size[e]	Minimum customer tail size				Minimum size of earthing conductor	Minimum size of bonding conductor
				Note[a]	Note[b]	Note[c]	Note[d]		
125KVA	200	160/200	95mm^2	70mm^2	95mm^2	50mm^2	70mm^2	35mm^2	25mm^2
150KVA	**200A**	**200A**	**95mm^2**	70mm^2	95mm^2	50mm^2	**70mm^2**	**35mm^2**	**25mm^2**
235KVA	400A	315A	150mm^2	150mm^2	240mm^2	95mm^2	150mm^2	70mm^2	35mm^2

[a]Table 4D1A – Cu Method C PVC Clipped direct.
[b]Table 4D1A – Cu Method B PVC in trunking.
[c]Table 4E1A – Cu Method C XLPE Clipped direct.
[d]Table 4E1A – Cu Method C XLPE in trunking.
[e]Concentric cable aluminium phase conductors.

6.5 SELECTING THE CPD SIZES

The CPD sizes are generally selected by calculation based on the design current. In the case of the sub-mains cable, the CPD size also needs to be determined by assessing the expected maximum demand and the peak current, consideration being given to discrimination of the downstream devices.

The design current (I_b) is determined by the load connected to the circuit. In is the rated current of the CPD (or the current setting on an adjustable device).

From the design process described in Chapter 4, it will be recalled that the rating of the CPD (I_n) must be greater or equal to the design current (I_b) so that it doesn't operate under normal conditions. The current-carrying capacity of the cable (I_z) must be greater than the rating of the CPD (I_n) so:

$$I_z \geq I_n \geq I_b$$

In this example the office power distribution board has a calculated current of 17.9A per Phase, so the minimum device size must be greater than 17.9A. The next standard CPD size is a 20A device, therefore it would be possible to use a 20A CPD. However, to allow discrimination with the largest downstream devices and give the end user more flexibility a 63A device would be selected as the minimum CPD for sub-mains.

In the case of the offices power DB, the largest CPD would be a 32A device downstream, and so a 63A CPD should provide adequate selectivity and discrimination with the downstream CPD protecting the final circuit cabling. This is shown in Fig. 6.3 below. The red curve is the DNO 200A BS88 Fuse, the green is the 63A MCCB supplying the office power DB, and the blue curve is the 32A Type C MCB supplying the largest outgoing way, any larger outgoing way would not provide discrimination with the CPD supplying the sub-main.

In certain cases consideration may be required as to the type of protection to be employed, for example, electronic trip units may be needed to gain discrimination. Each Warehouse Power DB requires a 109A supply to a 250A rated DB. This allows more flexibility by having a greater number of higher rated outgoing ways. In any case, the tenant may not split the load 50/50 between the two DBs. A 160A incoming device could be used (restricting the 250A DB to 160A maximum incomer), but it is also required that the larger incoming device discriminates with both the upstream 200A BS88 supply and the largest possible outgoing device, in this case a 63A MCB. From Fig. 6.3 it can be seen that a standard 160A Thermal magnetic MCCB (the green curve) will not discriminate with the 200A CPD of the DNO upstream. However, by using an electronic trip unit which has more flexibility, the MCCB can discriminate with both the 200A BS88 and the largest outgoing way (63A Type C MCB) (Fig. 6.4).

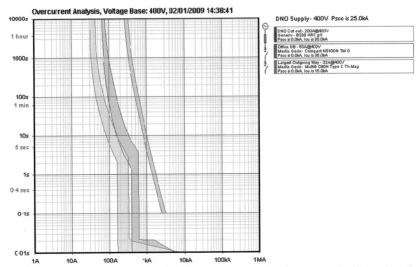

FIGURE. 6.3 Time/current characteristics of example circuit (office power distribution board) circuit protective devices to illustrate their co-ordination and discrimination (Amtech).

A similar process is followed for each of the other sub-main circuits to determine the CPD for each. In the chosen example circuit, the warehouse lighting DB has been selected and this is supplied *via* a 40A TPN MCCB, selected as the next available size above the design current of 36.6A/phase.

6.6 SELECT THE CABLE TYPE AND INSTALLATION METHOD

In this installation, the the sub-main cabling would generally be Multicore 90 °C Armoured cabling with thermosetting insulation and copper conductors, with the insulation being LSF (Cu XLPE/SWA/LSF). Refer to Chapter 12 of this book to see the benefits of this cable. The benefits of using this type of cable are relevant in this worked example as higher conductor ratings are afforded by the use of XLPE insulation, the cables which are exposed in the warehouse are protected by the SWA and the life safety aspects are enhanced by the LSF insulation. The data for this cable can be found in IEE Table 4E4A, which gives the relevant cable rating data.

The cabling could be installed on horizontal perforated cable tray, which is installation method No. 31 in IEE Table 4A2. This relates to reference method E or F for the current-carrying capacity in IEE Table 4E4A. The tabulated ratings of the cables are therefore found in either Column 4 or 5 depending on if it's single or three-phase supply, respectively.

From this table, the initial cable size can be determined based on the CPD rating (*In*). In the example the warehouse lighting DB has a rated *In* of 40A,

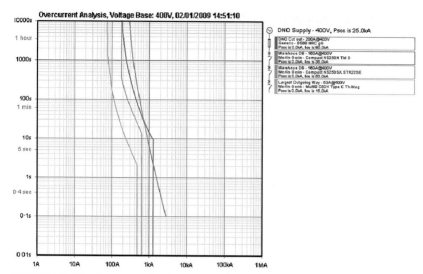

FIGURE 6.4 Time/current characteristics of warehouse power distribution board circuit and its circuit protective devices to illustrate their co-ordination and discrimination (Amtech).

which would equate to a 4.0-mm^2 cable. The actual tabulated current-carrying capacity of this cable (I_t) is 44A.

Rating Factors

As described in the earlier design section, rating factors will need to be applied to correct the conditions of the normal load, in accordance with the formula:

$$I_t \geq \frac{I_n}{C_a \times C_g \times C_i \times C_c}$$

Looking at each rating factor in turn:

C_a – Ambient temperature. The design criteria state that the ambient temperature will be 30 °C, therefore from IEE Table 4B1, which details the rating factors for ambient air temperature for this installation's situation, the sub-mains the correction factor will always be 30 °C so C_a is 1.00.

C_g – Grouping. The grouping is given in IEE Tables 4C1–4C5. There are a number of ways to calculate the grouping factors, which depend on how they are installed, the comparative loadings of the circuits, the arrangement of the cables and the phasing (i.e. SPN/TPN).

In this example, consider IEE Table 4C1, which gives the rating factors for one circuit/one multi-core cable/group of circuits/group of multi-core cables to be used with IEE Tables 4D1A–4J4A.

Firstly determine how many cables will be in the group. In the worst case this will occur as they split from the main panel where seven circuits will run together towards the end of the warehouse. If the cables are multi-core, installed in a single layer on horizontal cable tray, the initial grouping factor will equate to 0.73.

In the IEE Grouping tables there are a number of footnotes and these will need to be considered to cover cases where loading or cable types are not uniform. Where the conductors are lightly loaded (circuits carry less than 30% of its grouped rating), they may be discounted from the rating factor. This is applied with the following formula:

$$0.3 \times I_t \times C_g$$

In this example, there are seven circuits, so C_g equals to 0.73. Taking the Warehouse lighting DB supply, $I_t = 44A$ so $0.3 \times 44 \times 0.73 = 9.636$. As the load is greater than this figure it must be included, which would be the same for the circuits supplying DB1 and 2, DB5, and the ventilation panel.

For the FACP and the Sec Control Panel, the design current is $<6.351A$ $(0.3 \times 29 \times 0.73)$ so that these can be excluded. This means that the grouping factor is to be applied for five cables, not the seven on the tray. Thus a factor of 0.75 is to be applied.

With respect to the other rating factors, such as C_c and C_i, no insulation or rewirable fuses are used; therefore a factor of 1 can be used for these.

Once the rating factors have been determined, they can be applied to the cable ratings, for the example, cable supplying the warehouse lighting:

$$I_t \geq \frac{I_n}{C_a \times C_g \times C_i \times C_c} = I_t \geq \frac{40}{1 \times 0.75 \times 1 \times 1} = 53.33A$$

Therefore the tabulated cable rating (I_t) must be equal or greater than 53.33A, which the initially selected cable is not (44A). Therefore referring back to IEE Table 4E4A, a 6mm^2 cable will need to be specified (56A).

6.7 VOLTAGE DROP

Now that the initial cable selections based on the Design current Ib, the rating factors and actual cable tabulated ratings have been made, it is necessary to determine whether the cables selected are within the parameters set for allowable voltage drop (VD) within the distribution system.

Appendix 12 of the IEE Regulations indicates that the following VDs are to be adhered to:

Lighting circuits	3%
Other	5%

In the example of the sub-main cabling, it is necessary to also consider the final circuit cabling from the distribution boards to ensure that sufficient allowance remains.

It is usual to split the allowance 50/50, i.e. 2.5% for the sub-main and 2.5% for the final circuits, but the individual circumstances of each cable also need to be considered. For instance, the warehouse circuits will have a longer final circuit length than the office circuits. So a balance needs to be sought.

To calculate the basic voltage drop on a circuit the following information is required:

- The Length of the cable – this information was previously given in Table 6.8.
- The Design Current (Ib) – this has been previously calculated.
- The VD per A per metre (mV/A/m) of the cable to be used.

The VD in mV/A/m can be found in Appendix 4 of the IEE Regulations in the table corresponding to the current-carrying capacities of the relevant cable.

For this example, IEE Table 4E4B, Column 4 is used.

The following formula then gives the actual VD.

$$VD = \frac{(mV/A/m \times L \times I_b)}{1000}$$

For the warehouse lighting circuit, the components are:

Length	40m
Design current	36.6A
mV/A/m	6.8(6-mm^2 cable)

Therefore,

$$VD = \frac{(6.8 \times 40 \times 36.6)}{1000} = 9.96V = 2.49\% \text{ of } 400V$$

This is within the specified limits, but will leave only 0.51% for the final circuits (note, 3% maximum VD for lighting). This is lower than the acceptable figure and therefore a larger size should be considered. A 10mm^2 cable (4.0mV/A/m) would give a VD of 5.856V or 1.46% of 400V, leaving 1.54% for the final circuits. However, as the final circuits will be quite long, it may be more economical to increase the size of the sub-main cable and to reduce the size of the multiple cables that make up the final circuits. By selecting a 16mm^2 cable (2.5mV/A/m) the VD is 3.66V or 0.92% of 400V leaving 2.08% for the final circuits. This same method would be applied to the other sub-main cables.

Note that the calculation above is based on the design current. If the CPD rating had been used (I_n) the maximum current figure would be higher and the circuit may be outside the parameters for volt drop. Thus any future additions will need to take the increase in the above parameters into account.

6.8 PROSPECTIVE FAULT CURRENTS

Short-Circuit Current

The maximum and minimum short-circuit current will need to be established, the former is to enable equipment with the correct withstand capacity to be selected and the latter is to enable circuit to be checked for disconnection. The necessary information to complete the calculations can be obtained from cable and CPD manufacturers.

The maximum Prospective Short-Circuit Current (PSCC) quoted by the DNO will be the worst case and will account for any changes on their network that may occur. In a three-phase system the maximum value of fault level is determined by taking the impedance of one phase only up to the point being considered and then dividing it into the phase to neutral (nominal to earth) voltage.

$$I_{FMAX} = U_o/Z_s$$

where I_{FMax} is the fault current (maximum) and U_o is the nominal line voltage to earth (230V).

When calculating the values for shock protection, the external earth loop figure can be utilised.

$$Z_s = Z_e + R_1 + R_2$$

where Z_s is the earth fault loop impedance, Z_e is the external earth fault loop impedance, R_1 is the resistance of the phase conductor from the origin of the circuit and R_2 is the resistance of the earth conductor from the origin of the circuit.

The Z_e quoted as 0.2ohms by the DNO (Section 6.4) is the worst case minimum impedance and this would equate to a fault current of 1.15kA. In this example, the actual prospective external earth loop impedance in relation to the maximum fault level quoted ensures that equipment is specified which can handle the highest fault. The maximum PSCC at the supply terminals has been quoted by the DNO as 16kA, and from this the external Z_s can be calculated by U_o/I_{FMax} which equates to 0.0144ohms. A diagrammatic representation of the figures for the supply, sub-main and final circuits is given in Table 6.15.

The supply details are provided by the DNO, and it is known that the maximum fault level at point A (Table 6.15) will be 16kA. This is well within the capability of the HRC fuses protecting the main panel, but the characteristics at the load end of the DNO cables (point C) are to be calculated.

As previously stated, the DNO cables are 70mm^2 Cu/PVC [IEE Table 4D1A], for the phase conductor and 35mm^2 Cu/PVC for the main earthing conductor. From the data in Table 6.13, conductor resistances of 0.268 and 0.524ohms/km are derived, assuming a maximum cable length of 5m.

At point C (Table 6.15) the minimum value of loop impedance Z_{smin}

An R_1 value of 5m $\times$ 0.268/1000 = 0.00134ohms

An R_2 value of 5m $\times$ 0.524/1000 = 0.00262ohms

Therefore $Z_s = 0.0144 + 0.00134 + 0.00262 = 0.0184$ohms, and

$I_{FMAX} = U_o/Z_s = 230/0.0184 = 12,500$A or 12.5kA

maximum fault level.

This is well within the capability of the MCCBs protecting sub-main cabling supplied from the main panel. These calculations are carried out using the 20°C cable data, which provides the value when the circuit is first energised, representing lowest resistance and therefore the highest fault current.

TABLE 6.13 Cooper resistances for Copper Conductors (Draka)

Conductor area (mm^2)	Solid Conductor (Class 1) copper (ohms)	Stranded Conductor (Class 2) copper (ohms)	Flexible Conductor (Classes 5 and 6) copper (ohms)
1.5	12.1	12.1	13.3
2.5	7.41	7.41	7.98
4	4.61	4.61	4.95
6	3.08	3.08	3.30
10	1.83	1.83	1.91
16	1.15	1.15	1.21
25	0.727	0.727	0.78
35	0.524	0.524	0.554
50	0.387	0.387	0.386
70	0.268	0.268	0.272
95	0.193	0.193	0.206
120	0.153	0.153	0.161
150	0.124	0.124	0.129
185	–	0.0991	0.106
240	–	0.0754	0.081

Next, the minimum prospective fault current must be considered. This is required to ensure that the circuit is disconnected in the required time. Again, this takes the worst case situation, which will be at the load end of the circuit, calculated under the normal operating temperature of the conductor.

The conductor resistances used above need to be adjusted for the normal operating temperature of the conductor by applying a factor to the $20\,°C$ conductor resistance. This factor is determined by taking the difference in the temperature from $20\,°C$ to the operating temperature and multiplying by the resistance coefficient of the material.

The copper conductor coefficient of resistance at $20\,°C$ is 0.004, the temperature difference between the operation and $20\,°C$ is 50 $°C$ ($70°C$ − $20\,°C$). Thus 0.004×50 equals to 0.20 and, as this is to be added to the resistance of the conductor, a factor of 1.20 can be used.

At point C, the maximum value of loop impedance Z_{smax}

An R_1 value of 5m $\times$ 0.268 $\times$ 1.2/1000 = 0.00161ohms
An R_2 value of 5m $\times$ 0.524 $\times$ 1.2/1000 = 0.00314ohms
Therefore Z_s = 0.0144 + 0.00161 + 0.00314 = 0.0192ohms, and
I_{FMAX} = U_o/Z_s = 230/0.0192 = 11,979 A minimum fault level.

IEE Table 41.4 states that the maximum Z_s value for a 200A BS88 fuse (i.e. the DNO supply) is 0.019ohms, therefore this meets the regulation. Note that IEE Fig. 3.3A states that a minimum fault level of 1200A is required to achieve a 5s disconnection time, which this easily achieves, and as the value is over 3000A the protective device will disconnect in less than 0.1s.

From the main panel (point C), taking the example of the sub-main cable to the warehouse lighting DB, the value of the conductor resistances must be added to those already calculated for the supply cables. The circuit is supplied by 40m of $16mm^2$ Cu XLPE/SWA/LSF cable, the values of which are given in Table 6.14, as well as the details of the CPC which are to be provided by the SWA of the cable.

To check whether the protective device will disconnect in the required time, the maximum value of Z_s is required, and by using the data from Table 6.14, conductor resistances of 1.466 and 3.1ohms/km are obtained for the phase and earthing conductors, respectively.

As stated, this needs to be adjusted for the normal operating temperature of the conductor by applying a factor to the $20\,°C$ temperature conductor resistance. This factor has been previously determined for the phase conductor as 1.2 and for the SWA, the coefficient of resistance at $20\,°C$ is 0.0045, and the assumed operating temperature is $60\,°C$. This gives 0.18, therefore the factor of 1.18 is to be used.

Note that although the operating temperature of XLPE is 90 $°C$, as it is not fully loaded a 70 $°C$ value has been taken, which is the same as the PVC value above.

TABLE 6.14 Extract of Data for Four-Core Cu/XLPE/SWA/LSF Cables to BS 5467 (Draka)

Nominal area of conductor (mm²)	Approx. overall diameter (mm)	Approx. cable weight (kg/km)	Maximum resistance of cable ac at 20°C (ohms/km)	Reactance @ 50Hz (ohms/km)	Impedance @ 90°C AC (ohms/km)	Maximum armour resistance @ 20°C	Gross CSA of armour wires (mm²)
1.5	13.5	365	15.4280	0.104	15.428	8.80	17
2.5	15.0	438	9.4480	0.101	9.449	7.70	20
4	16.4	532	5.8780	0.099	5.879	6.80	22
6	18.7	764	3.9270	0.094	3.928	4.30	36
10	21.1	1013	2.3330	0.093	2.336	3.70	42
16	22.9	1360	1.4660	0.088	1.469	3.10	50
25	27.6	2160	0.9260	0.082	0.930	2.30	70
35	30.4	2690	0.6685	0.077	0.673	2.00	78

TABLE 6.15 Results of Sub-Main Cable Calculations

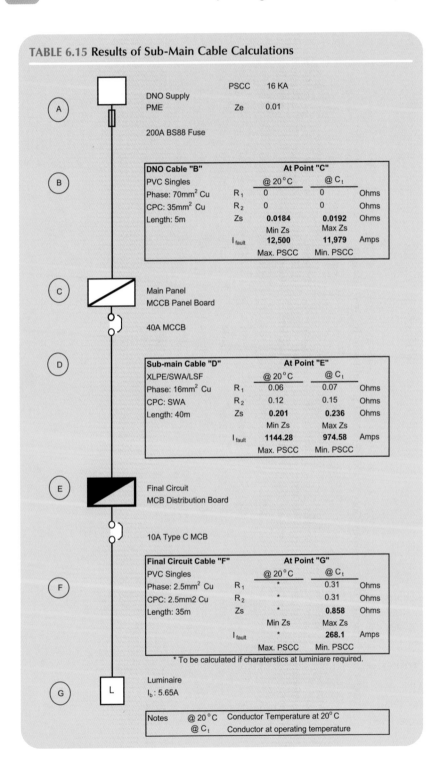

A

DNO Supply
PME

200A BS88 Fuse

PSCC	16 KA	
Ze	0.01	

B

DNO Cable "B"		At Point "C"		
PVC Singles		@ 20°C	@ C_1	
Phase: 70mm² Cu	R_1	0	0	Ohms
CPC: 35mm² Cu	R_2	0	0	Ohms
Length: 5m	Zs	**0.0184**	**0.0192**	Ohms
		Min Zs	Max Zs	
	I_{fault}	**12,500**	**11,979**	Amps
		Max. PSCC	Min. PSCC	

C

Main Panel
MCCB Panel Board

40A MCCB

D

Sub-main Cable "D"		At Point "E"		
XLPE/SWA/LSF		@ 20°C	@ C_1	
Phase: 16mm² Cu	R_1	0.06	0.07	Ohms
CPC: SWA	R_2	0.12	0.15	Ohms
Length: 40m	Zs	**0.201**	**0.236**	Ohms
		Min Zs	Max Zs	
	I_{fault}	**1144.28**	**974.58**	Amps
		Max. PSCC	Min. PSCC	

E

Final Circuit
MCB Distribution Board

10A Type C MCB

F

Final Circuit Cable "F"		At Point "G"		
PVC Singles		@ 20°C	@ C_1	
Phase: 2.5mm² Cu	R_1	*	0.31	Ohms
CPC: 2.5mm2 Cu	R_2	*	0.31	Ohms
Length: 35m	Zs	*	**0.858**	Ohms
		Min Zs	Max Zs	
	I_{fault}	*	**268.1**	Amps
		Max. PSCC	Min. PSCC	

* To be calculated if charaterstics at luminiare required.

G

Luminaire
I_b: 5.65A

Notes	@ 20°C	Conductor Temperature at 20°C
	@ C_1	Conductor at operating temperature

This provides:

An R_1 value of 40m $\times$ 1.466 $\times$ 1.2/1000 $=$ 0.0704ohms and
An R_2 value of 40m $\times$ 3.1 $\times$ 1.18/1000 $=$ 0.1463ohms
Therefore $Z_{smax} =$ 0.0192 $+$ 0.0704 $+$ 0.1463 $=$ 0.236ohms, and
$I_{FMAX} = U_o/Z_s =$ 230/0.236 $=$ 974.58A minimum fault level.

Protection Against Electric Shock

To protect against electric shock *via* either direct contact (basic protection) or indirect contact (fault protection), a number of protective measures are available. Automatic Disconnection of the Supply (ADS) is achieved by insulating the live parts to provide the basic protection and automatically disconnecting the supply *via* protective bonding to provide fault protection. The Steel Wire Armouring (SWA) of the cabling which also has both inner and outer insulations provides other protective measures.

To check that the fault protection requirement is met the maximum disconnection time stated within the IEE Regulations must not be exceeded. This time is derived from the current that will flow under fault conditions and the time the CPD takes to operate to disconnect the circuit. This can be determined by referring to the time/current characteristics found in Fig. 6.5, and finding the corresponding current that is required to operate the device. The maximum earth fault loop impedance can then be determined using the following formula:

$$Z_s \geq U_o/I_f$$

The IEE Regulations tabulate the results of this formula for many types of protective device. However, some are not included and in these cases manufacturer's data will need to be sought. In our example, the MCCB falls into this category and therefore information will need to be sourced from the manufacturer to determine the characteristics of the device.

As the example circuit is a distribution circuit and a TN system, IEE Regulation 411.3.2.3 allows a maximum disconnection time of 5s. It has been found that the maximum earth fault loop impedance at point E was calculated to be 0.236ohms (with the minimum earth fault current) using the $Z_s \leq U_o/I_f$ formula above. From the manufacturer's data, the maximum earth loop impedance is 0.58ohms, therefore the circuit shall disconnect within the required time.

Referring to the time/current graph in Fig. 6.5, it can also be seen that the 40A MCCB will actually disconnect instantaneously ($<$0.01s).

Protective Conductors

All conductors must be thermally protected, and this is usually the case for the phase and neutral conductors. However, it is quite common for the protective conductor to be of a different size or conducting material to that of

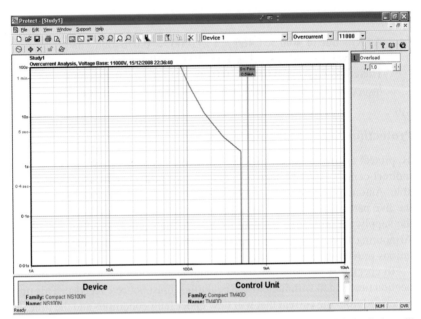

FIGURE 6.5 Time/current graph of a 40A MCCB showing that at the PSCC the device will actually disconnect in less than 0.01s.

the phase conductors and therefore it is necessary to confirm that the protective conductors are thermally protected. There are two approaches to this. In one, the conductor is selected using IEE Table 54.7. Otherwise they must comply with the adiabatic equation, which will give a more accurate requirement.

The adiabatic equation can be found under IEE Regulation 543.1.3, and is as follows:

$$S = \frac{\sqrt{I^2 t}}{k}$$

where S is the cross-section area of the conductor (mm^2); I is the fault current (A); t is the duration of the fault current (s); and k is the conductor material factor, based on resistivity, temperature coefficient and heat capacity.

Values of k are to be found in IEE Tables 54.2–54.6 and are dependent on the installation method and conductor material.

For the example circuit, it has already been found that the maximum fault current is 12.5kA and the duration is 0.01s. The k value needs to be selected from the IEE Tables, and since the SWA is being utilised for the CPC, IEE Table 54.4 is to be used.

Using steel for the conductor material and 90 °C thermosetting for insulation material, the corresponding value of k is 46.

Using these figures in the equation provides:

$$s = \frac{\sqrt{12500^2 \times 0.01}}{46} = 27.2\text{mm}^2$$

Table 6.14 shows that the equivalent armouring CSA for a 16mm^2 four-core SWA cable is 50mm^2 and so the armour is thermally protected.

If the circuit were to fail this criterion, two options are open. The first is to increase the size of feeder cable so that the earth loop impedance meets the requirement. The second is to install a separate protective conductor to reduce the earth loop impedance. It should be noted that the IEE Regulations do not recommend using the SWA which is then supplemented by a separate CPC to make up the requirement for the protective device as it cannot be guaranteed that the current would be shared proportionally between the two different conductors due to the different resistive properties of the materials, and the magnetic field associated with the armour.

6.9 CONTAINMENT SIZING

The sub-main routing has been decided along with the installation method and now that the sizes of the sub-main cables are known, the containment size can be assessed. From experience, the likely containment size will be known, but it will be necessary to determine the actual minimum size by calculation. For this installation, medium duty perforated cable tray has been chosen which is supplied in fixed sizes, the type and duty will be dependent on the application and the spans to be covered.

Unlike conduit and trunking there are no prescribed methods for sizing cable tray, but one method may be to determine the physical size of the cabling and prepare a layout of an indicative section. Due regard must be taken for cable fixing methods such as cleating or tie-wrapping, together with extra spacing for segregation for heat dissipation and an allowance for future provision.

An example of this could be taking a section as the cables run along the south wall of the warehouse, as indicated with Fig. 6.1. Assuming that the same process above has been carried out for the other sub-main cabling, the resultant cables sizes may be as listed in Table 6.16. The outside diameters of each of the cables can be found from manufacturer's data, such as those listed in Table 6.14. It can be seen that allowing for suitable fixings and spacing, the cables require 185.6mm across the section of the containment, and allowing for 25% additional space for future capacity, this equates to 232mm of space required. This is just over a 225mm tray and therefore next available standard manufacturer's tray size of 300mm must be selected. This will achieve 38% extra space for future cabling.

TABLE 6.16 Cable Detail Sizes Through Containment Section

Circuit reference	Supplying	Cable size	Cable type	Cable diameter/ space required (mm)
Space for clipping				5
1/L1	Office lighting distribution board	Three-core 10mm²	Cu PVC/XLPE/ SWA/LSF	19.5
Space for clipping			5mm	
1/L2	Office power distribution board	Three-core 10mm²	Cu PVC/XLPE/ SWA/LSF	19.5
Space for clipping				5
1/L3	Fire detection and alarm control panel (FACP)	Three-core 4mm²	Cu PVC/XLPE/ SWA/LSF	15.3
Space for clipping				5
2	Warehouse lighting distribution board	Four-core 16mm²	Cu PVC/XLPE/ SWA/LSF	22.9
Space for clipping			5mm	
4	Warehouse Tenant Fit-out Distribution Board No. 2	Four-core 50mm²	Cu PVC/XLPE/ SWA/LSF	32
Space for clipping				5
5	Ventilation/ mechanical services control panel	Four-core 10mm²	Cu PVC/XLPE/ SWA/LSF	21.1
Space for clipping				5
6/L1	Security Control Panel	3 Core 4mm²	Cu PVC/XLPE/ SWA/LSF	15.3
Space for clipping				5
Total Space Required for cables				185.6
Including 25% allowance for future cabling				232

Return Flange - Medium Duty

Straight Tray

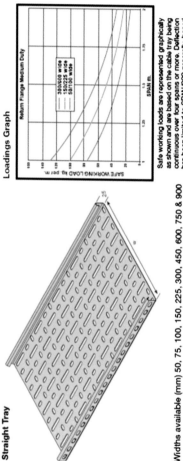

Widths available (mm) 50, 75, 100, 150, 225, 300, 450, 600, 750 & 900
All widths have slots in sides
Standard length - 3m

Identification and Selection

Loadings Graph

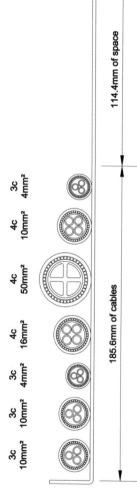

Safe working loads are represented graphically as shown and are based on the cable tray being continuous over four spans or more. Deflection has been limited to SPAN/200 generally, based on the end span condition as the worst case. Deflection will be less than this on internal spans. However, on wider trays, additional deflection will be induced locally across the base of the tray, depending on the width of the tray and the load distribution across the width. This will not be detrimental to the structural performance of the tray but may need consideration if appearance is of prime importance.

Use with containment section

| 3c 10mm² | 3c 10mm² | 3c 4mm² | 4c 16mm² | 4c 50mm² | 4c 10mm² | 3c 4mm² |

← 185.6mm of cables → | ← 114.4mm of space →

FIGURE 6.6   Containment section showing example cable arrangement.

To demonstrate, these cables could be laid out as indicated within Fig. 6.6. This process can be repeated at each intersection, or periodically along the containment routes as cables leave the containment to ensure that the correct containment has been selected and also that the containment has not been unnecessarily oversized.

6.10 FINAL CIRCUITS

Next, the final circuits emanating from the DBs can be considered. It is possible to make initial assessments of the final circuit arrangements prior to this point, but having carried out the sub-main distribution design it is possible to gain a more accurate assessment of the source characteristics and available VDs at the final DBs.

It can be appreciated that the number of final circuits in an installation such as this will be quite large and for simplicity just a single final circuit will be considered. In this case it will be a final circuit of the warehouse lighting system. The design process is similar to that for the sub-main cabling, and therefore it will be worked through in a simplified form. Of course, as before, the actual loading of the circuit will need to be determined.

The estimated load imposed on the DB was calculated to be 20.3kW, 36.6A per phase, based on the 'area × W/m^2' figure, but to determine the actual final circuit load a lighting design will need to be conducted.

Lighting Design

It is outside the scope of this text to detail a complete final lighting design process, but an estimation is shown, based on the criteria available. One method to determine the amount of lighting required is by using the Lumen method, and there are a number of steps involved.

Firstly, a suitable luminaire is selected and, considering the design parameters in Table 6.4, a low-bay discharge luminaire is required. There are a number of options available, but typically for this installation a 250W, HQI-T metal halide lamp could be chosen, as detailed within Figs 6.7–6.9.

Secondly, the Room index (K) is required, this being the relationship between the proportions of the room.

$$K = \frac{L \times W}{(L + W) \times H_m}$$

where K is the room index; L is the length of the room; W is the width of the room; and H_m is the height of luminaire above working plane.

And from the data presented previously within this chapter,

Area of warehouse	$1350m^2$
Dimensions	45m $(L) \times$ 30m $(W) \times$ 5.5m (H)[a]
Working plane	0m (floor)
Height of luminaire	4.0m as detailed in design criteria
Therefore H	4.0m

[a]Assumed that warehouse takes up the whole area, i.e. including the office and amenity for simplicity (this would be refined in the actual final scheme).

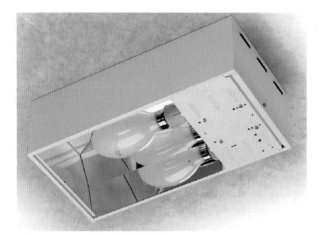

FIGURE 6.7 A typical low-bay luminaire c/w a 250W, HQI-T metal halide lamp source (Cooper Lighting & Security).

PHOTOMETRIC DATA

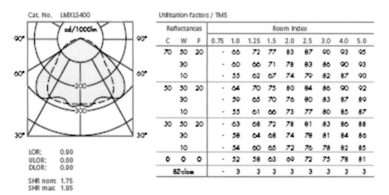

Cat. No. LMXLS400

Utilisation factors / TMS

Reflectances			Room Index								
C	W	F	0.75	1.0	1.25	1.5	2.0	2.5	3.0	4.0	5.0
70	50	20	·	66	72	77	83	87	90	93	95
	30		·	60	66	71	78	83	86	90	93
	10		·	55	62	67	74	79	82	87	90
50	50	20	·	64	70	75	80	84	86	90	92
	30		·	59	65	70	76	80	83	87	89
	10		·	55	61	66	73	77	80	85	87
30	50	20	·	63	68	72	78	81	83	86	88
	30		·	58	64	68	74	78	81	84	86
	10		·	54	60	65	72	76	78	82	85
0	0	0	·	52	58	63	69	72	75	78	81
BZ-class			·	3	3	3	3	3	3	3	3

cd/1000lm

LOR: 0.90
ULOR: 0.00
DLOR: 0.90

SHR nom: 1.75
SHR max: 1.95

FIGURE 6.8 Typical photometric data for the luminaire proposed (Cooper Lighting & Security).

METAL HALIDE Tubular clear lamp							
Designation	Watts (W)	Nominal Dimensions (mm)	Cap	Colour Temp (K)	Initial Lumens (lm)	Rated Life (50% Survivors)	Lumen Maintenance At Rated Life
HQI-T	250	Dia 46 x L 225	E40	4200	21000	10000 hr	62%

FIGURE 6.9   Typical lamp details for the luminaire proposed (Cooper Lighting & Security).

Therefore,

$$K = \frac{45 \times 30}{(45 + 30) \times 4} = 4.5$$

Thirdly, the required number of luminaires is determined, this uses the formula:

$$N = \frac{E \times A}{F \times \text{MF} \times \text{UF}} = \frac{300 \times 1350}{21,000 \times 1 \times 0.8 \times 0.94} = 25.65$$

or approximately 26 luminaires required, where N is the number of luminaries required; E is the lighting level required – 300 Lux; Area 1350m² (45m $(L) \times$ 30m (W)), F 21,000 – initial bare lamp lumens from lamp table (Fig. 6.9); MF 0.8 – see notes below; and UF 0.94 – see notes below.

Note

MF is the Maintenance factor where a figure of 0.8 has been assumed. The actual figure is based on the lamp lumen factor (the reduction in the lumen output after a specific number of operation hours), the lamp survival factor (the percentage of lamp failures), and the lamp and room maintenance factor (the reduction of light due to dirt on the lamp and the environment), based on BS EN 12464.

UF is the utilisation factor, which by using the K (room index) value obtained, can be taken from the utilisation factor tables, the data of which is shown within Fig. 6.8. The data is given for specific room reflectances, therefore with a room index of about 4.5 and assuming average ceiling, wall and floor reflectances of 70, 50, 20, respectively, the UF from Table 6.9 is 0.94 (between 4.0–(93) and 5.0–(95)).

Once the required number of luminaires to provide the required lighting level is known, they can be laid out in the space to provide a uniform illumination. As 26 does not divide easily into a grid, a suitable arrangement would be four rows of seven luminaires, giving a total of 28. To check the layout, these luminaires should be laid out to ensure that a uniform distribution of light and the luminaries should not exceed the manufacturer's recommended space to height ratios (SHR).

For this example, the worst case is the shortest side, four fittings in 30m. This gives 7.5m maximum spacing and with a 4m height the ratio is 1.875. It can be seen from the photometric data in Fig. 6.8 that the nominal SHR is 1.75, with a maximum figure of 1.95. As 1.875 falls between these figures, this should

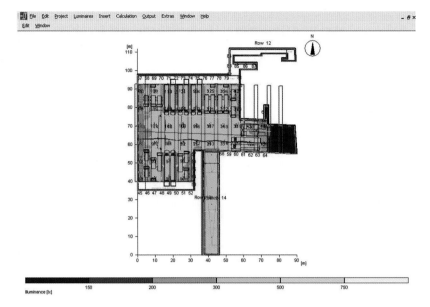

FIGURE 6.10 Typical example of computer generated lighting calculation (Relux).

provide an acceptable uniformity. It can therefore be confirmed that the lighting layout will consist of four rows of seven luminaires to achieve the required lighting levels, which can be checked by reworking the lumen method formula, as shown below.

$$E = \frac{F \times n \times \text{MF} \times \text{UF}}{A} = \frac{21{,}000 \times 28 \times 0.8 \times 0.94}{1350} = 327.5 \text{ Lux Average}$$

The final lighting design uses the actual luminaire photometric data and a point by point calculation which is usually completed using software packages, an illustration of which is shown in Fig. 6.10.

Circuiting

Now that the quantity of fittings is known, a more detailed circuit design may be carried out so that the design current of the circuit can be established. Generally, each row will be fed *via* a separate circuit but the number of circuits per row needs to be determined. Manufacturer's data can be used to determine how many fittings can be accommodated on to a single circuit protective device, an example is shown in Table 6.17.

This example would require a maximum of five fittings for a 10A type C MCB therefore with seven fittings per row, it would be necessary to split the circuit into three and four fittings.

TABLE 6.17 Maximum Quantities of Discharge Luminaires Per MCB (Cooper Lighting & Security)

Lamp power and type	C10 MCB	C16 MCB	C20 MCB
50W SON	19	31	39
50W MBF	16	24	31
70W SON & HQI	12	18	23
150W SON & HQI	7	11	14
250W SON & HQI	5	7	9
250W MBF	4	6	7
400W SON & HQI	3	4	5
400W MBF	2	4	5

Determine the Design Current

Now that the circuits are known, the design current can be determined. This is a simple exercise to sum the loads supplied by the circuit. For this example the circuit that contains four fittings will be used.

From the electrical data previously shown in Fig. 6.2, each fitting has a running VA of 325 @ 230V = 1.413A, that is 5.652A per circuit and 12A starting current.

The total circuit power figures are given, which include the ballast, control gear and lamp wattages for the luminaire. This data can be used to work back to a W/m^2 figure as follows:

$$28 \times 276 = 7728/1350 = 5.72 \text{W}/\text{m}^2$$

This is about 1/3rd of the expected design criteria, based on a 'rule of thumb' for the lighting level requirements. A more accurate initial assessment would be the use of a W/m^2/100 Lux figure, but over the site as a whole the 15W/m^2 figure should average out as it may be greater in the office and reception areas where a higher light level is required.

Selecting the CPD Sizes

It is known from the manufacturer's data that a maximum of five luminaries can be supplied by a 10A Type C MCB, but the actual quantity of luminaires needs to be investigated to ensure that unintentional operation (nuisance tripping) of the CPDs doesn't occur. This would be carried out by checking the peak inrush starting currents against the minimum tripping current of the circuit breaker.

Selecting the Cable Type and Installation Method

The warehouse lighting is to be mounted on galvanised steel trunking, suspended from the roof structure. For this type of installation it would be common to expect the circuits to be wired in Cu LSF single cables, therefore the cable details will be as detailed in IEE Table 4D1A and to installation method 10, reference method B of IEE Table 4A2.

Determine the Initial Cable Size

Now the design current and CPD size is known, the initial size of the final circuit conductor can be determined.

$$I_t \geq \frac{I_n}{C_a \times C_g \times C_i \times C_c}$$

where $I_n = 10A$ CPD; $C_a = 1.00$ as the Ambient temperature is the 30 °C as the sub-main example; $C_g = 0.52$ see note below; $C_c = 1.0$ the circuit isn't backed up by a rewirable fuse therefore a factor of 1 can be used; $C_i = 1.0$ the cabling doesn't pass through any insulation therefore a factor of 1 can be used.

Grouping

Each row of luminaires will be supplied by two circuits, which split into trunking carrying just one circuit each. Thus the majority of the circuits will not be grouped at this stage with any other cables. Cables will be grouped where they leave the DB, and it is known that there are eight circuits in total, together with a number of other final circuits. Details and loadings of all these circuits are currently not known, although it could be assumed that a number will be lightly loaded and therefore discounted from the group. In the initial case eight circuits will be assumed, although when further details are known, the design can be revisited and checked. From IEE Table 4C1, reference method B, the grouping factor is 0.52. Once the rating factors have been determined, they can be applied to the cable ratings, for our example:

$$I_t \geq 10/(1 \times 1 \times 0.52 \times 1) = 19.23A$$

Therefore the tabulated cable rating (I_t) must be equal or greater than 19.23A, referring back to IEE Table 4D1A, a 2.5mm^2 cable will need to be specified (I_t 24A).

Voltage Drop

The VD calculations are carried out as before and for this example the final circuit to the luminaires will be *via* the trunking, with the luminaires attached underneath. Assuming a final circuit length of 35m including the cable drop to the DB, and as it has been seen, a design current of 5.65A.

The tabulated VD is derived from IEE Table 4D1B, and the VD of a 2.5mm^2 cable is 18mV/A/m.

Using the VD formula:

$$VD = \frac{(mV/A/m \times L \times I_b)}{1000}$$

For the warehouse lighting circuit, the components are:

Length	3 5m
Design current	5.65A
mV/A/m	18
= 3.56V = 1.55% of 230V	

As the sub-main volt drop has already been calculated, it is known that 2.08% VD remains for the final circuits. Therefore the use of a 2.5mm^2 cable complies as the total VD will be 2.47%, which is below the 3% limit requirement.

Determination of Supply Characteristics

With further reference to Table 6.15, the minimum expected PSCC and maximum external earth loop impedance for the distribution board have already been determined and this information can be used for the supply characteristics of the final lighting circuit (at point E).

The minimum fault level on the sub-main to the warehouse DB (point E) is the fault level at the end of the circuit, which is also the start of the final circuits from the DB. This figure was 975.2A, based on a Z_s of 0.236ohms. These figures can be used to calculate the maximum Z_s at point G, but firstly the maximum PSCC at point E is to be calculated to ensure that the CPD protecting cable 'F' can withstand the prospective fault current at that point in the circuit.

Prospective Fault Currents – Short-Circuit Current

To calculate the maximum short-circuit current, the same conductor resistances for the sub-main cable supplying the DB can be utilised but they must be taken at 20 °C, i.e. without the temperature correction, hence:

$R_1 = 40 \text{ m} \times 1.466/1000 = 0.059\text{ohms, and}$
$R_2 = 40 \text{ m} \times 3.1/1000 = 0.124\text{ohms}$
Therefore $Z_s = 0.0184 + 0.059 + 0.124 = 0.201\text{ohms, and}$
$I_{FMAX} = U_o/Z_s = 230/0.201 = 1144.28\text{A or } 11.4\text{kA maximum fault level.}$

This is well within the capability of the MCB protective device protecting the final circuit, which has a withstand value of 15kA.

As before, the minimum prospective fault current is calculated at point 'G' which involves taking the conductor resistances of the example final circuit and adding them to external loop impedance of the circuit which was previously calculated (0.236ohms). The final circuit consists of 35m of 2.5mm² Cu LSF single cables, which has a conductor resistance of 7.41ohms/km, taken from Table 6.13. Assuming that the CPC is of the same size as the phase conductor, the values will be:

$$R_1 \text{ value of } 35 \times 7.41 \times 1.2/1000 = 0.311\text{ohms and}$$
$$R_2 \text{ value of } 35 \times 7.41 \times 1.2/1000 = 0.311\text{ohms}$$

Therefore,

$$Z_s = Z_e + R_1 + R_2 = 0.236 + 0.311 + 0.311 = 0.858\text{ohms, and}$$
$$I_{FMin} = U_o/Z_s = 230/0.858 = 268A \text{ minimum fault level.}$$

Protection Against Electric Shock

As with the sub-main calculations above, the circuit must be checked to ensure that the protective device will disconnect within the required time. In the example our final circuit is rated at less than 32A with a U_o of 230V, therefore the required disconnection time from IEE Table 41.3 is 0.4s. Unlike the sub-main circuit considered earlier, the results of this formula have been tabulated in the IEE Regulations for this type of CPD against the required disconnection time.

Referring to IEE Table 3.3, a 10A type C circuit breaker would require a maximum earth fault impedance of 2.3ohms, so therefore it should disconnect with the required time of 0.4s. The time/current graph in IEE Table 3.5 shows that the 10A Type C MCB will actually disconnect in >0.1s.

Protective Conductors

As before, to determine that the protective conductors are thermally protected and of the correct size the adiabatic equation can be applied, this will be based on the following:

I, fault current	268.1A
T, duration of the fault current	0.1s
K, conductor material factor	115

The value of K is taken from IEE Table 54.3, as a separate Cu/PVC LSF cable is being used for the CPC.

Using these figures in the equation gives:

$$s = \frac{\sqrt{268.1^2 \times 0.1}}{115} = 0.74\text{mm}^2$$

Therefore the selected conductor size is thermally protected and shall carry the PSCC current effectively.

Finally

Calculations for sizing of circuits in conduit are given in Chapter 9, and similar methods may be applied where trunking is to be used.

As the design progresses, and a more precise estimation evolves for each DB, this can be used for updating the sub-mains and therefore the supply characteristics.

When the design is completed and final locations of accessories, sockets, plant and lighting are known, the design calculations can be revisited and updated with the most current information. This process will have many iterations and is the normal next follow-on step in the design process.

In this worked example, only a few circuits have been analysed and calculated in detail. However, the calculations for each of the other circuits in any installation would be carried out in a similar way to those illustrated, the task being made more speedily with the use of modern design software.

Chapter 7

Special Types of Installation

Certain types of installation demand special consideration when designing and installing the electrical equipment. Part 7 of the 17th edition of the IEE Wiring Regulations sets out the specific needs of some types of special installation and the IEE Regulations contained therein supplement or modify the other parts of the IEE Regulations. These types of special installations are:

- Locations containing a bath or shower (701)
- Swimming pools and other basins (702)
- Rooms and cabins containing sauna heaters (703)
- Construction and demolition site installations (704)
- Agricultural and horticultural premises (705)
- Conducting locations with restricted movement (706)
- Electrical installations in caravan/camping parks and similar locations (708)
- Marinas and similar locations (709)
- Medical locations (710)
- Exhibitions, shows and stands (711)
- Solar photovoltaic (pv) power supply systems (712)
- Mobile or transportable units (717)
- Electrical installations in caravans and motor caravans (721)
- Temporary electrical installations for structures, amusement devices and booths at fairgrounds, amusement parks and circuses (740)
- Floor and ceiling heating systems (753)

The numbering of this section of the regulations is not sequential. The number appearing after the section number, e.g. 701.<u>32</u> generally refers to the corresponding part of the regulations from Part 1 to 6, i.e. Chapter 32.

Not all of these installations will be part of the day-to-day work of a designer or electrician, and so the detail described below covers only the more prevalent types encountered.

Two of the special types of installation which were covered in the 16th edition of the IEE Regulations, are now dealt with differently. The section on installations with high protective conductor currents is now covered by IEE Regulation 543.7 of the regulations and highway power supplies are in IEE Regulation 559.10.

7.1 LOCATIONS CONTAINING A BATH OR SHOWER

In rooms containing a bath or shower the risk of electric shock is increased due to the fact that the body is in contact with earth and, as a result of being wet, has reduced electrical body resistance. IEE Regulations Section 701 details the requirements and specifies zones in bath and shower rooms with restrictions on equipment which may be fitted.

Additional protection must be provided for all circuits by the use of RCDs and limitations exist as to the forms of protection which may be used. Supplementary equipotential is generally provided which connects the terminals of all the protective conductors of circuits for class I and II equipments together, along with all accessible extraneous-conductive parts. This supplementary equipotential bonding may be omitted if certain conditions are met although the practicalities and verification of the conditions, and the need for peace of mind, mean it may be prudent to provide equipotential bonding anyway.

Section 701 describes which, if any, accessories and current-using equipment can be installed into which zones. Generally no switches, socket outlets or other electrical equipments may be installed unless certain conditions are met. Switches should be placed such that they are inaccessible to a person in the bath or shower, unless they are supplied by SELV (extra-low voltage) not exceeding 12V a.c. rms (or 30V d.c.), or are part of a shaver unit incorporating an isolating transformer to BS EN 61558. Similarly with socket outlets, none are permitted unless supplied by SELV, not exceeding 12V or are 3m away from the boundary of zone 1.

FIGURE 7.1 Cord-operated switches for use in bathrooms can be obtained in a variety of designs, with or without a neon pilot light and, if needed, with engraved labelling as with the fan isolator shown.

Cord operated switches may be used provided the switch itself is located within the correct zone, also where SELV or PELV is used, the equipment must be provided with basic insulation, and be protected to IPXXB or IP2X.

7.2 SWIMMING POOLS AND OTHER BASINS AND ROOMS CONTAINING A SAUNA

As with bath and shower rooms, increased precautions against electric shock are required in these locations and in certain specified zones within or near them. IEE Regulations Sections 702 and 703 set out the details for swimming pools and sauna heaters, respectively. Requirements include the provision of barriers with appropriate degrees of protection in accordance with BS EN 60529, placing certain equipments outside specified zones, provision of SELV (extra-low voltage) supplies, protective measures such as the use of residual current devices (RCDs) and equipotential bonding and constraints on the type of wiring systems which may be used. In the case of hot air saunas, provision to avoid the overheating of electrical equipment must be made.

There is a duty upon the designer to extend the assessment of general characteristics. Specific examination of these areas and the way in which they may be used must be made. Additional information is contained in the IEE Regulations themselves and in the IEE books of Guidance Notes.

7.3 CONSTRUCTION AND DEMOLITION SITE INSTALLATIONS

Temporary electrical installations on building and construction sites are necessary to enable lighting and power to be provided for the various trades engaged on the site. These temporary installations need to be of a very high standard owing to the exceptional hazards which can exist.

The Electricity at Work Regulations 1989, which applies to permanent installations, also apply to temporary installations on construction sites, so these temporary installations must be of the same standard as those for other installations. Installations are also required to comply with British Standard BS 7375 and the IEE Regulations Section 704 also applies.

The old practice of using brass lamp holders with twisted two-core flexible cord was the cause of many accidents. The use of these in the vicinity of earthed metal or damp floors presented a real hazard, even when connected to extra-low voltage supplies. Apart from the danger of shock there is a danger to the eyes should the lamp be accidentally broken. All portable hand lamps must be properly insulated and fitted with a guard.

Construction site lighting is necessary to cover the following requirements:

1. Lighting of working areas, especially internal working areas where there is no natural light, with a minimum intensity of 20 lux. In cases where

FIGURE 7.2   Load centre for a temporary installation on a construction site (William Steward & Co).

activities include more detailed types of work such as brick or slab laying, the minimum intensity recommended is 50 lux.

2. Walkways, especially where there are uneven floors, minimum intensity 5 lux.
3. Escape lighting, along escape routes, this lighting to be from a supply separate from the mains supply, usually battery operated, minimum intensity 5 lux.
4. Emergency lighting. This to be in accordance with BS 5266 Part 1 and to come on automatically in the event of mains failure. Usually battery operated or from a generating set. Minimum intensity 5 lux.

High-level fixed lighting could be taken from 230V mains supply, but low level and portable lighting should be 110V with centre point earthed *via* a double-wound transformer.

Power for tower cranes, mixers and other motors over 2kW is usually supplied from a 400V mains supply. Sockets for portable tools and hand lamps should be to BS EN 60439-4, and it is strongly recommended that they are fed from a 110V supply *via* a double-wound transformer. In vulnerable situations, such as in damp areas, tanks etc., the voltage should be reduced to the SELV voltage (50V a.c.).

IEE Regulations Section 704 includes a number of additional and amended regulations which apply to construction sites. Because of the increased risk of hazards which exist in these locations some tightening of requirements, particularly with regard to shock protection, is called for.

A number of BS (British Standards) apply to installations on construction sites. BS 7375 and BS EN 60439-4 cover electricity supplies and equipment,

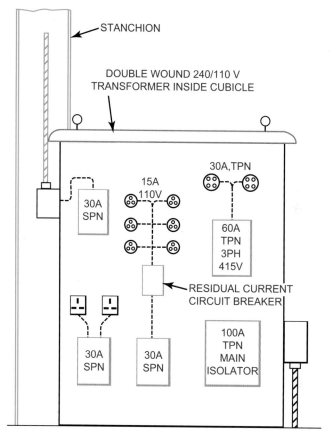

FIGURE 7.3 Diagrammatic view of the load centre for a temporary installation.

BS EN 60309-2 applies to plugs, sockets and cable couplers, and certain minimum standards for enclosures are required consistent with BS EN 60529: 1992.

Electrical conductors must not be routed across roadways without adequate mechanical protection, and all electrical circuits must have isolators at each supply point which are capable of being locked in the off position. Additional useful information is contained in Section 704 of the IEE Regulations.

7.4 AGRICULTURAL AND HORTICULTURAL PREMISES

Agricultural installations, which include buildings accessible to livestock, require very special consideration. Horses and cattle have a very low body resistance which makes them susceptible to electric shock at voltages lower than 25 V a.c. The IEE Regulations include a number of requirements specific to these applications and these include revised arrangements for Automatic disconnection of supply, use of extra-low voltage circuits, Supplementary bonding, Accessibility and Selection and erection of wiring.

It is recommended that electrical equipment installed in these areas should have a degree of protection to at least IPXXB or IP2X. Switches and other accessories should be placed out of reach of animals and this generally means that they be placed in enclosures or outside the areas occupied by livestock. In the case of low-voltage systems, the circuits should be protected by a residual current circuit breaker, and for socket outlets this must have an operating current not exceeding 30mA.

As with other areas of high risk from shock currents, modified arrangements are a requirement of the IEE Regulations regarding the times for automatic disconnection and other associated measures. Supplementary equipotential bonding connecting all exposed and extraneous conductive parts must be provided, and this includes any conductive or metal mesh covered floors. Bonding conductors must be mechanically protected and not subject to corrosion.

In some cases, supplies are needed for life support of livestock, and for these separate final circuits must be provided, with an appropriate alternative back-up supply.

Mains-operated electric fence controllers must comply with BS EN 60335-2-76 and BS EN 6100-1 but their installation is not covered by the IEE Regulations.

7.5 ELECTRICAL INSTALLATIONS IN CARAVAN PARKS, CARAVANS AND MOTOR CARAVANS

Under the 17th edition of the IEE Regulations, Caravan and Camping Parks are covered in Section 708 and Caravans and Motor Caravans in Section 721.

Definitions
- The IEE Regulations define a caravan park is an area of land that contains two or more caravan pitches and/or tents.
- A caravan as 'a trailer leisure accommodation vehicle, used for touring, designed to meet the requirements for the construction and use of road vehicles'. The IEE Regulations also contain definitions of motor caravan and leisure accommodation vehicles.

Caravan Parks

As electrical installations on caravan sites are extremely vulnerable to possibilities of shock due to their temporary nature, special regulations have been made. A socket outlet controlled by a switch or circuit breaker protected by an overcurrent device and a RCD shall be installed external to the caravan, and shall be enclosed in a waterproof enclosure (min. IP44), it shall be non-reversible with provision for earthing BS EN 60309-2.

External installations on caravan parks, although some may only be of a temporary nature, must be carried out strictly to the IEE Regulations in general, and the Electricity Supply Regulations. The preferred method of supply is by underground distribution circuit and the depth of this and degree of mechanical protection are specified. In the case where overhead distribution is to be used, the conductors must be insulated and a minimum height of 3.5m is specified, or 6.0m where vehicle movements could take place.

Caravan and Motor Caravan Installations

All mobile and motor caravans shall receive their electrical supply by means of a socket outlet and plug of at least 16 A capacity, with provision for earthing BS EN 60309-2. These sockets and plugs should have the phase, neutral and earth terminals clearly marked, and should be sited on the outside of the caravan. They should be connected to the main switch inside the caravan by cables 25m (±2m) in length. A notice must be fixed near the main switch inside the caravan bearing indelible characters, with the text as given in IEE Regulation 721.514.1. This notice gives instructions to the caravan occupier as to precautions which are necessary when connecting and disconnecting the caravan to the supply. It also recommends that the electrical installation in the caravan should be inspected and tested at least once every three years, and annually if the caravan is used frequently.

Other recommendations are that all wiring shall be insulated single-core cables installed in non-metallic conduit or sheathed flexible cables. Cables shall be firmly secured by non-corrosive clips at intervals not exceeding 250mm for horizontal runs and 400mm for vertical runs. Luminaires shall be mounted direct on the structure; flexible pendants must be suitable for the movement of the vehicle. Protective conductors should be incorporated in the cable containing the circuit conductors or their conduit.

Where automatic disconnection of supply is used, an RCD is to be provided, and the protective conductor arrangements are specified. These must terminate at an earthing terminal connected to the structural metalwork which is connected to the protective contacts of socket outlets, exposed-conductive-parts of electrical equipment and connected to the earthing contact of the caravan inlet.

7.6 MARINAS AND SIMILAR LOCATIONS

This new part of the IEE Regulations, Section 709, is not covered in detail by this text, but has similar requirements to those of caravan parks. The treatment of temporary connection of pleasure crafts and house boats is dealt with in a similar way, but with the added external influences of water, corrosive elements, movement of structures, i.e. boats bobbing around and the presence of fuel etc. Diagramatic means of obtaining electricity supplies in Marinas are included in the IEE Regulations.

7.7 MEDICAL LOCATIONS

This is a new section within the IEE Regulations, but is reserved for future use and therefore there are currently no requirements stated within the 2008 edition of the IEE Regulations.

7.8 SOLAR PHOTOVOLTAIC (PV) POWER SUPPLY SYSTEMS

An increasing trend to consider on-site electrical generation from a renewable source is covered by a new section of the IEE Regulations. This specifies requirements for both the d.c. element of current generation and the a.c. side where an a.c. module is provided. The preferred method of protection for the d.c. side is by the use of Class II or equivalent insulation. The installation method is to ensure that there is adequate heat dissipation to cope with maximum solar radiation conditions.

Isolation arrangements are specified as are the increased requirements arising due to external influences such as wind, ice formation and solar radiation.

Other special types of installation which are not covered by this text include:

- Conducting locations with restricted movement (706)
- Exhibitions, shows and stands (711)
- Mobile or transportable units (717)
- Temporary electrical installations for structures, amusement devices and booths at fairgrounds, amusement parks and circuses (740)
- Floor and ceiling heating systems (753)

Further information for the sections not covered may be obtained by reference to Section 7 of the IEE Regulations and the IEE Books of Guidance Notes.

7.9 OTHER SPECIAL INSTALLATIONS

There are other installations and systems that are not specifically covered by Section 7 of the IEE Wiring Regulations, but which are worthy of consideration. Some of these are covered below.

Emergency Supplies to Premises

The need for emergency supplies in factories, commercial buildings, hospitals, public buildings, hotels, multi-storey flats and similar premises is determined by the fire prevention officer of the local authority concerned, and is also related to the need to provide a minimum level of continuity of supply.

In large installations, it is usual to provide standby diesel driven alternators for essential services. It is unlikely these will be able to supply the full load of the building, and thus some load shedding will be necessary. Because the occupants of the building will not normally carry out this function, some special circuit provision should be made. The essential loads should ideally be determined at the design stage, and separate distribution arrangements made from the main switchboard. The changeover switching arrangements are usually automatic, and the circuits must be arranged in such a way that the standby power supplies feed only the essential load distribution network.

Wiring to emergency supplies must comply with the IEE Regulations and BS 5266. Recommended systems of wiring are MI cables, PVC/armoured cables, FP cables, PVC or elastomer insulated cables in conduit or trunking. In certain installations the use of plastics conduit or trunking is prohibited, and the enforcing authority should be consulted on this.

Emergency Escape Lighting

The object of emergency lighting is to provide adequate illumination along escape routes within 5 s of the failure of normal lighting. BS 5266 deals with the emergency lighting of premises other than cinemas and certain other premises used for entertainment.

If the recommendations of BS 5266 Part 1 are complied with it is almost certain that the emergency lighting system will be acceptable to the local 'enforcing authority'.

The Fire Protection Act of 1971 indicates the need for escape lighting, but does not make any specific demands. IEE Regulation 313.2 mentions that any emergency supplies required by the enforcing authority should have adequate capacity and rating for the operation specified. BS 5266 Part 1 recommends that emergency lighting be provided in the following positions:

1. Along all escape routes towards and through all final exits, including external lighting outside all exits.
2. At each intersection of corridors, and at each change of direction.
3. On staircases to illuminate each flight of stairs, and near any change of floor level.
4. To illuminate all exit signs, directional signs, fire alarm contacts and fire fighting appliances. (Note: The illumination of signs may be either from within or external to the sign.)
5. All lifts in which passengers may travel.
6. All toilet areas which exceed eight square metres.
7. Over moving staircases or walkways (i.e. escalators and travelators) as if they were part of the escape route.
8. Control, plant, switch and lift rooms.

Emergency lighting must come into operation within 5s of the failure of the normal lighting, and must be capable of being maintained for a period from 1 to 3 hours (according to the requirements of the local 'enforcing authority'). The level of illumination should be not less than 0.2 lux measured at floor level on the centre line of the escape route.

FIGURE 7.4 Automatic emergency lighting unit. The unit has a self-contained rechargeable battery, and control gear which detects mains failure and energises the emergency light.

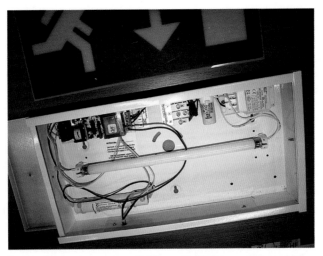

FIGURE 7.5 The Automatic emergency lighting unit with the cover removed. The control gear can be seen and the rechargeable battery located at the lower left hand side.

Along corridors it is recommended that the spacing of lighting luminaires should have a maximum ratio of 40:1 (i.e. distance between luminaires and mounting height above floor level) and, of course the illumination levels above must also be met.

Alternative methods of providing emergency lighting are as follows:

1. Engine driven generating plants, capable of being brought on load within 5s.

2. Battery powered systems, utilising rechargeable secondary batteries, combined with charger, centrally located to serve all emergency lights.
3. Signs or luminaires with self-contained secondary batteries and charger. The battery after its designed period of discharge must be capable of being re-charged within a period of 24h.

Circuits feeding luminaires or signs with self-contained batteries shall be continuously energised, and steps must be taken to ensure that the supply is not inadvertently interrupted at any time. Switches or isolators controlling these and other emergency lighting circuits must be placed in positions inaccessible to unauthorised persons, and suitably identified.

All wiring for emergency lighting and fire alarms if enclosed in conduit or trunking must be segregated from all other wiring systems (see IEE Regulation 528-01-04). When trunking is used the emergency lighting and fire alarm/ circuits, cables must be segregated from all other cables by a continuous partition of non-combustible material.

Multi-core cables should not be used to serve both emergency and normal lightings (BS 5266).

Standby Supplies

In addition to emergency escape lighting, it is very often desirable to provide 'standby supplies' which will come into operation in the event of a failure of the supply. This lighting is intended to provide sufficient illumination to enable normal work to be carried on. It is very often necessary where there are continuous processes, which must not be interrupted, and in computer installations. In these cases it is also necessary to provide standby power supplies to enable the processes to continue.

Fire Alarms

The design of fire alarm systems does not come within the scope of this book, and it is usual for manufacturers of fire alarm equipment or other specialists in this work to design these installations.

Fire alarm systems are covered by British Standard BS 5839. Generally speaking the approved systems of wiring are the same as those for emergency lighting. Wiring installed in conduits or trunking must be segregated from all other types of wiring systems (except emergency lighting).

Installations in Hazardous Areas

Hazardous areas mainly consist of places where potentially flammable materials are present. This includes spraying and other painting processes which involve the use of highly flammable liquids, locations where explosive dust

FIGURE 7.6   Fire alarm point and siren wired in MI cable.

may be present, installations associated with petrol service pumps, and inspection pits in garages.

Electricity at Work Regulation 6 states that electrical equipment which may reasonably foreseeably be exposed to hazardous conditions 'shall be of such construction or as necessary protected as to prevent, so far as is reasonably practicable, danger arising from such exposure'.

The Fire Offices Committee has issued recommendations for electrical installations in connection with highly flammable liquids used in paint spraying. Conditions also exist for the granting of petroleum spirit licences in respect to electrical equipment. These conditions require that petrol pumps shall be of flameproof construction, so also shall be switchgear and other electrical control gear. Luminaires within the pump equipment shall be of flameproof construction, but those mounted outside the pump casing shall be of totally enclosed design in which the lamp is protected by a well glass or other glass sealed to the body of the luminaire so as to resist the entry of petroleum spirit vapour. The wiring shall be carried out by insulated cables enclosed in heavy gauge galvanised solid-drawn steel conduit. Conduit boxes within the pump equipment shall be of flameproof construction and galvanised. Alternative wiring may consist of MI cables, copper sheathed with flameproof glands.

The supply circuits for each pump shall be separately protected with over-current protection, and these protective devices shall not be situated within, or on, the pump housing. BS EN 60079-1 gives details of the flameproof enclosures.

Where explosive dusts are likely to be present, flameproof equipment and circuit systems must be used, but the luminaires and conduit fittings and other electrical equipment must also be fitted with dust-tight gaskets to prevent the entry of explosive dusts. Without these dust-tight gaskets the ordinary flame-proof accessory could breathe in explosive dusts between the machined surfaces when changes in temperatures occur.

Part II

Practical Work

Chapter 8

A Survey of Installation Methods

The preceding chapters cover the various regulations governing the control and distribution of supplies, and the design and planning of installation work have been discussed. From now on the practical aspects of electrical installation work will be dealt with. It is very important that the practical work is carried out correctly.

8.1 CABLE MANAGEMENT SYSTEMS

Commercial and industrial electrical installations are generally comprehensive and complex systems, and when installed in new or recently refurbished buildings employ a range of methods in the distributing and routeing of electrical circuits. A number of firms are able to supply the full range of equipment needed. This includes a variety of trunking and conduit types, cable tray, cable ladder and such other items as power poles or posts, with which the electrical installer can present a complete and well-finished installation.

The collective term used for the variety of methods available is 'cable management systems' and the various elements of them are dealt with separately in this book under the appropriate chapters on conduit, trunking or busbar asystems.

Principal Types of Wiring Systems

There are many alternative wiring systems that may be adopted:

1. PVC (thermoplastic) single-core insulated cables (70 °C) in conduit, ducts or trunking;
2. Rubber (thermosetting) insulated cables (90 °C) in conduit, ducts or trunking;
3. PVC insulated and sheathed multi-core (flat) cables fixed to a surface or concealed;
4. MI or FP cables;
5. PVC single-core sheathed on cleats;
6. PVC multi-core sheathed and armoured;
7. EPR or XLPE multi-core armoured;
8. PILC multi-core armoured (encountered in existing installations);
9. Busbar systems.

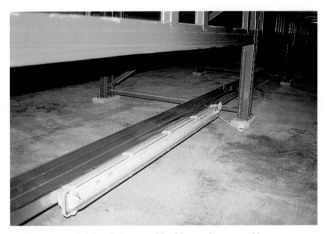

FIGURE 8.1   Emergency lighting fittings used in this warehouse stacking system are mounted at floor level on multi-compartment trunking. This enables separation of circuits of different voltage bands (W.T. Parker Ltd).

Installation Method

Single or multi-core cables run in conduit or trunking,
Larger sizes of cable run on cable tray or ladder,
Cables clipped direct to a surface,
Cables run in cable basket,
Armoured cable,
Busbar System.

Cable types could be:
70 °C Thermoplastic (e.g. PVC),
90 °C Thermosetting (e.g. XLPE or EPR),
CWZ (MI, FP) BS6387,
Chemical resistant.

Choice of Wiring System

In deciding the type of wiring system for a particular installation, many factors have to be taken into consideration; amongst these are:

1. whether the wiring is to be installed during the construction of a building, in a completed building, or as an extension of an existing system;
2. capital outlay required;
3. planned duration of installation;
4. whether damp or other adverse conditions are likely to exist;
5. type of building;
6. usage of building;
7. likelihood of alterations and extensions being frequently required.

TABLE 8.1 Glossary of Terms used in Cable Identification

Acronym	Meaning	Notes
CPE	Chlorinated polyethylene	
CSP	Chlorosulphonated polyethylene	
EPR	Ethylene propylene rubber	
ETFE	Ethylene tetrafluoroethylene	
FP	Fire performance	
FR	Fire retarded	
HOFR	Heat oil resistant and flame retardant	
LDPE	Low-density polyethylene	
LSF	Low smoke and fumes	Made to BS 6724
LSOH	Low smoke, zero halogen	No halogens in cable insulation
MI	Mineral insulated	See Chapter 14
NH	Non-halogenated	Made to BS 6724
PILC	Paper insulated lead covered	Sometimes with PVC sheath
PTFE	Polytetrafluoroethylene	
PVC	Polyvinylchloride	Widely used. Suitable for 70 °C. Made to BS 6346
RS	Reduced smoke	
SWA	Steel wire armoured	
Tri-rated	PVC insulated cables for switchgear and control wiring complying with three standards:-	
	(1) Type CK cables to BS 6231	
	(2) Type TEW equipment wires to Canadian Standard C22.2 No. 127	
	(3) American Underwriters Laboratories (UL) Subject 758	
VR	Vulcanised rubber	
XLPE	Cross-linked polyethylene	Suitable for 90 °C. Made to BS 5467

FIGURE 8.2   PVC/SWA/PVC sheathed cables being used as sub-main cables for distribution of power. Provision has been made for 12 240mm² four-core cables to be run, and in this view 11 have been fitted (William Steward & Co Ltd).

TABLE 8.2 Maximum Normal Operating Temperatures Specified for the Determination of Current-carrying Capacities of Conductors and Cables

Type of insulation	Maximum or sheath normal operating temperature, °C
Thermoplastic (PVC) compounds	70
MI cables exposed to touch or plastic covered	70
Thermosetting (90 °C rubber) compounds	90
LSOH and heavy duty HOFR compounds	90
XLPE to BS 5467	90
Bare MI cables not exposed to touch and not in contact with combustible material	105
Varnished glass fibre	180

Note: Where the insulation and sheath are of different materials, the appropriate limits of temperature for both materials must be observed.

Since the relative importance of each of the foregoing factors will vary in each case, the final responsibility must rest with the person planning the installation.

Frequently a combination of several wiring systems may be used to advantage in any one installation. For example, in an industrial installation the main and sub-main cables would probably consist of XLPE/SWA/PVC cables, or MI copper sheathed cables. The power circuits could use LSF, XLPE or PVC insulated cables in conduit or trunking, or MI cable. The lighting circuits could be carried out with PVC cables in plastic trunking or conduit, or with PVC insulated and sheathed cables fixed to the surface.

Low-Voltage Wiring

The 17th edition of the IEE Regulations deals with two voltage ranges under the definition 'nominal' voltage:

Extra-low voltage	not exceeding 50V a.c. or 120V ripple-free d.c. whether between conductors or to earth.
Low voltage	exceeding extra-low voltage but not exceeding 1000V a.c. or 1500V d.c. between conductors or 600V a.c. or 900V d.c. between any conductor and earth.

All conductors shall be so insulated and where necessary further effectively protected be so placed and safeguarded as to prevent danger so far as is reasonably practicable.

FIGURE 8.3 This neat installation shows cable tray with PVC and FP cable systems used to good advantage in a distribution area.

8.2 FOUNDATIONS OF GOOD INSTALLATION WORK

Whatever wiring system is employed there are a number of requirements and regulations which have a general application. These will be mentioned before dealing with the various wiring systems in detail. Regulation 134.1.1 in Chapter 13 of the IEE Regulations says 'Good workmanship … and proper materials shall be used …'. This is no mere platitude because bad workmanship would result in an unsatisfactory and even a dangerous installation, even if all the other regulations were complied with. Good workmanship is only possible after proper training and practical experience. Knowledge of theory is very necessary and extremely important, but skill can only be acquired by practice. It must always be remembered that the choice of materials, layout of the work, skill and experience all combine to determine the character and efficiency of the installation.

Proper Tools

There is an old adage about being able to judge a workman by his tools, this is very true. An electrician must possess a good set of tools if the work is to be carried out efficiently. Moreover, the way in which the electrician looks after his/her tools and the condition in which they are kept is a very sure indication of the class of work likely to be produced. Besides ordinary hand tools, which by custom electricians provide for themselves, there are others such as stocks and dies, some types of power tools, bending machines, electric screwing machines which are usually provided by the employer, and which will all contribute to the objective of 'good workmanship'.

FIGURE 8.4 A high-level cable route in an industrial building is illustrated in this view. Cable ladder is provided to carry distribution cables feeding switchboards remote from the site sub-station. Cable basket, which in this case has three separate compartments, is to be used for data cabling and cable tray is provided for final circuits feeding lighting and machinery (W.T. Parker Ltd).

Selection of Cable Runs

Cables and other conductors should be so located that they are not subject to deterioration from mechanical damage, vibration, moisture, corrosive liquids, oil and heat. Where such conditions cannot be avoided, a suitable wiring system must be provided. Four examples are: (1) MI cables which will withstand water, steam, oil or extreme temperature, (2) LSF or PVC covered MI cables which will withstand, in addition, chemicals corrosive to copper, (3) XLPE up to 90 °C and (4) PVC/SWA/PVC sheathed cables will withstand most of these conditions and operating temperatures up to 70 °C. Table 8.2 gives the maximum normal operating temperatures of various types of cables.

If operating temperatures exceeding 150 °C are encountered then special heat resisting cables must be used, such as varnished glass fibre, for temperatures up to 250 °C, and where there are exceptionally high temperatures the conductors must be of high melting point materials, such as nickel or chromium copper, or silver plated copper.

It should be noted, as explained in Chapter 2, that the current rating of cables depends very much upon the ambient temperature in which they are installed. In boiler houses and similar installations where the cables are connected to the thermostats, immersion heaters and other equipment located near or on the boiler, it is usual to carry out most of the wiring with PVC cables in conduit or trunking, or with MI cables, and make the final connection near the boiler by means of a short length of suitable thermosetting insulated cable in flexible conduit, these cables being joined with a fixed connector block in a conduit box fitted a short distance away from the high-temperature area. The flexible metallic conduit permits the removal of the thermostat or other device without the need to disconnect the cables (Fig. 8.5).

Cables Exposed to Corrosive Liquids

Where cables are installed in positions which are exposed to acids or alkalis, it is usual to install PVC insulated cables. The use of metal covering should be avoided. Similarly, in the vicinity of seawater, steel conduits or other systems employing ferrous metals should not be used.

Cables Exposed to Explosive Atmospheres

Where conductors are exposed to flammable surroundings, or explosive atmospheres, special precautions have to be taken. The Electricity at Work Regulations 1989 state that electrical equipment which may reasonably foreseeably be exposed to hazardous environments 'shall be of such construction or as necessary protected as to prevent, so far as is reasonably practicable, danger arising from such exposure'. Such installations should comply with BS EN 60079. The Petroleum (Consolidation) Act as amended by the Dangerous

FIGURE 8.5 Wiring systems in use in a boiler house. The main wiring is in steel conduit and short lengths of heat-resisting cable in flexible conduit are used for the connections in the vicinity of the boiler where the temperatures are high (M.W. Cripwell Ltd).

Substances and Explosive Atmospheres Regulations 2002 deals with installations for petrol service pumps and storage depots.

Preventing the Spread of Fire

There are a number of aspects to be considered and these range from considering fire alarm circuits, layout of escape routes to limiting the extent of fire and smoke. Part B of the Building Regulations lays down requirements.

When installing conduits, trunking or cables in any building a very necessary precaution is to avoid leaving holes or gaps in floors or walls which might assist the spread of fire. Vertical cable shafts or ducts could enable a fire to spread rapidly through the building. Any holes or slots, which have to be cut in floors or walls to enable cables to pass through, must be made good with incombustible material.

Vertical cable ducts or trunking must be internally fitted with non-ignitable fire barriers at each floor level. The slots or holes through which the conduits or trunking pass must be made good at each floor level. The internal non-ignitable barriers not only restrict the spread of fire, but also counteract the tendency for hot air to rise and collect at the top of a vertical duct.

When fitting fire barriers, it is important to select the correct fire stop material. The use of incorrect material may achieve the desired result in preventing spread of fire, but may cause an unacceptable level of thermal insulation to be applied to the cables. If this occurs, the cable rating needs to be reduced, as with any cable run in thermally insulating material (Fig. 8.6).

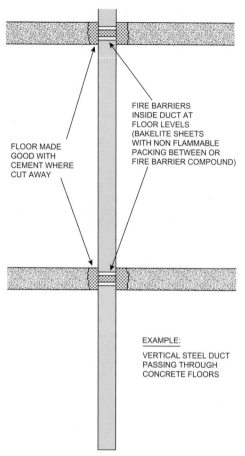

FIRE BARRIERS
INSIDE DUCT AT
FLOOR LEVELS
(BAKELITE SHEETS
WITH NON FLAMMABLE
PACKING BETWEEN OR
FIRE BARRIER COMPOUND)

FLOOR MADE
GOOD WITH
CEMENT WHERE
CUT AWAY

EXAMPLE:

VERTICAL STEEL DUCT
PASSING THROUGH
CONCRETE FLOORS

FIGURE 8.6 Preventing the spread of fire. In vertical cable duct fire barriers are fitted where the trunking passes through floors and the floors are made good with cement where cut away. Special fire barrier compounds are available which are elastomeric based and expand if exposed to high temperatures.

The Identification of Conduits and Cables in Buildings

IEE Regulation 514.1.2 demands that wiring is marked so that it can be identified during inspection, testing, repair or alteration. Harmonised cable colours are given in detail in Table 51 of the IEE Regulations. In essence, since harmonisation, colours for fixed cables are Brown, Black and Grey for phase conductors L1, L2 and L3, respectively, with Blue representing Neutral and Green/Yellow for Protective Conductors. In cases where alterations are to be made to an existing installation, it will at once be realised that since blue was formerly one of the phase colours, care will be needed when installing new wiring. It is a requirement for any installation using both the 'old' and present colour systems to display a warning notice.

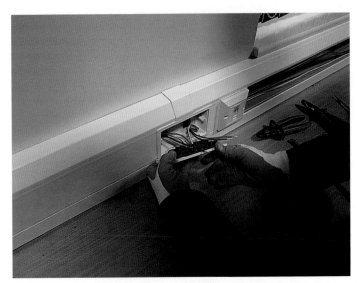

FIGURE 8.7   An extension to this installation will result in a mixture of colours from the old and current standards. A warning notice will need to be displayed (M.W. Cripwell Ltd).

FIGURE 8.8   In any installation where alterations to the wiring give rise to a mixture of colours from the old and current standards, it is a requirement to display a warning notice worded as shown. This should be placed 'at or near the appropriate distribution board' (IEE Regulation 514.14.1).

Conduit colours are specified and where it is practicable to paint them, they should be coloured orange. The incoming supply and the distribution cables should also be marked to show the nature of the supply, the number of phases and voltage.

Cables in Low-Temperature Areas

Some PVC insulated cables can be operated in temperatures down to $-30\,^{\circ}\text{C}$ and $-40\,^{\circ}\text{C}$ and manufacturers' data should be examined to check the

FIGURE 8.9 Identification of connections is essential and apart from compliance with regulations it ensures correct installation and assists with fault finding. These connections are in a control panel.

FIGURE 8.10 A number of types of cable idents are available. The type shown here clip onto the cable and can be clipped together in any combination.

minimum operating temperature of a particular cable type. However, cables should not be installed anywhere during periods when the temperature is below 0 °C, as the insulation is liable to crack if handled in very low temperatures, in fact the IEE On-site Guide recommends a minimum temperature of +5 °C during installation.

Single-Core Cables

Single-core cables, armoured with steel wire or tape shall not be used for a.c. circuits. However, single-core cables can sometimes be used with advantage and an example would be using, say, 630mm^2 cables between the supply transformer and the main control panel. In such a case, the use of Aluminium Wire Armouring (AWA) is perfectly acceptable.

Bunching of Outgoing and Return Cables

If the outgoing and return cables of a two-wire a.c. circuit, or all the phases and neutral of a three-phase circuit, are enclosed in the same conduit or armoured cable excessive induction losses will not occur.

Single-core cables (without armour) enclosed in conduits or trunking must be bunched so that the outgoing and return cables are enclosed in the same conduit or trunking. This must be accepted as a general rule for all a.c. circuits, and it must be ensured that no single conductor is surrounded by magnetic material, such as steel conduit, trunking or armouring. The reason is that any single-core cable carrying alternating current induces a current in the surrounding metal, which tends to oppose the passage of the original current.

If the outgoing and return cables are enclosed in the same conduit or trunking, then the current in the outgoing and return cables, each carrying equal current, will cancel each other out as far as induction is concerned, and therefore no adverse effects will occur. A voltage drop of 90%, and considerable overheating, has been known when single-core cables, enclosed separately in magnetic metal, have been connected to an a.c. supply.

Where it is essential those single-core cables are used in a particular application, and the protection of conduit or trunking is required, consideration should be given to the use of non-metallic enclosures. A number of plastic conduit and trunking systems are available.

There are occasions when the need to take precautions against induction is not observed. One example is when single-core cables enter busbar chambers, distribution boards or switchgear; if single-core cables carrying alternating currents enter these through separate holes in a metallic housing, circulating currents will be induced. In the past, manufacturers of switchgear and electric motors have provided three separate holes in the casing for three-phase circuits. If it is impossible for all the cables to pass through one hole then non-ferrous or aluminium gland plates must be used, or the space between the holes should be slotted (Fig. 8.11).

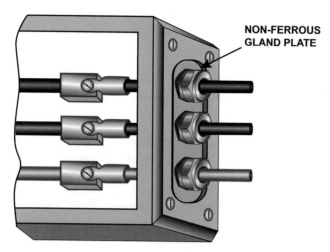

NON-FERROUS
GLAND PLATE

FIGURE 8.11 When single-core cables carrying heavy alternating currents pass through the metal casing of a switch, terminal box or similar equipment, they should, where possible, do so through a single hole; otherwise the space between the holes should (a) be slotted or (b) be fitted with a non-ferrous gland to prevent circulating currents.

Where SWA cable is to be used, circulating currents can be prevented by earthing only one end of the main cables. A separate cpc must, of course, be provided.

8.3 METHODS OF INSTALLATION

The IEE Regulations Appendix 4 and Tables 4A1 and 4A2 give details of various types of installations and these affect the current-carrying capacity of the cables. Installation methods covered in IEE Table 4A2 include clipped direct, embedded in building materials, installed in conduit, trunking or on cable trays and cables installed in enclosed trenches. The installation of cables where they are in contact with thermally insulating materials is also covered and the IEE Regulation relevant to this is 523.7. As mentioned in the Design section of this book, the current-carrying capacity of cables varies considerably according to the installation system chosen.

Cables with Aluminium Conductors

Multi-core sheathed cables with aluminium conductors are sometimes used instead of cables with copper conductors, as they are usually cheaper and are not so heavy as cables with copper conductors. IEE Tables 4H1A to 4H4B and 4J1A to 4J4B give current ratings of these cables, and it will be noted that the smallest size given in these tables is 16mm^2.

The current-carrying capacity of aluminium conductors is approximately 78% of the ratings for copper conductors, and therefore for a given current a larger cable will be necessary. The decision whether to use cables with copper or aluminium conductors will depend very much upon the market price of these two metals but of course other considerations must be taken into account, such as the fact that aluminium cables are generally of larger diameter than copper cables, and, as will be seen in the tables, the voltage drop is much greater (1.65 times that of copper).

The use of aluminium conductors presents some problems, but these can easily be overcome if the necessary precautions are taken. Aluminium, when exposed to air, quickly forms an oxide film which is a poor electrical conductor. If this film is allowed to remain it would set up a high resistance joint, and would cause overheating and eventually breakdown. There is also a risk, under damp conditions, of electrochemical action taking place between aluminium conductors and dissimilar metals. A further disadvantage is that the coefficient of expansion of aluminium is not the same as that for copper, and therefore terminations of aluminium conductors made to copper or brass terminals can give trouble if not properly made. Undoubtedly the best method of terminating aluminium conductors is to use crimping sockets made of tinned aluminium.

Before crimping, the conductors should be scraped to remove the oxide film, and then immediately smeared with a suitable paste such as 'Unial' to prevent further oxidation.

Segregation of Cables

Where cables are associated with extra-low voltage, fire alarm and telecommunications circuits, as well as circuits operating at low voltage and connected directly to a mains supply system, precautions shall be taken to prevent electrical contact between the cables of the various types of circuit. This is covered by IEE Regulation 528.1 and this regulation refers to the voltage bands I and II which are defined in Part 2 of the IEE Regulations. Band I broadly covers circuits such as telecommunications or communications circuits which are typically 'Extra-Low Voltage'. Band II covers 'Low Voltage', in no case exceeding 1000V a.c. The separation required can be achieved in a number of different means and these include cases where:

a. every cable is insulated for the highest voltage present;
b. each conductor of a multi-core cable is insulated for the highest voltage present in the cable;
c. each conductor of a multi-core cable is enclosed within an earthed metallic screen; or
d. cables are installed in separate compartments of a duct or trunking system.

Where cables enter a common box, circuits must be separated by partitions of fire-resisting material. Both metallic and non-metallic trunking systems are

FIGURE 8.12 Segregation of circuits is a requirement of the Regulations. In this view, cable basket, ready for erection, has been fitted with a dividing barrier (M.W. Cripwell Ltd).

available with suitable barriers which ease installation and enable the requirements to be met. An alternative means of segregation of different circuits would be to arrange for cables to be spaced apart for a sufficient distance.

The main object of these precautions is to ensure, in the case of fire, that alarm and emergency lighting cables are kept separate from other cables which might become damaged by the fire. BS 5266 deals with the segregation of these circuits and also gives details of other precautions which are necessary. Cables which are used to connect the battery chargers of self-contained emergency lighting luminaires to the normal mains circuits are not considered to be emergency lighting circuits.

Joints and Connections between Cables

Joints between cables should be avoided if possible, but if they are unavoidable they must be made either by means of suitable mechanical connectors or by soldered joints. In either case they must be mechanically and electrically sound and be readily accessible (IEE Regulations 132.12 and 526.3).

Electricity at Work Regulation 10 states that 'every joint and connection in a system shall be mechanically and electrically suitable for use'.

In some wiring systems, such as the MI and PVC sheathed wiring system, it is usual for the cables to be jointed where they branch off to lighting points and switches. These joints are made by means of specially designed joint boxes, or ceiling roses. In the conduit system it is usual for the cables to be looped from

FIGURE 8.13 Screwed steel conduit is a widely used and very effective cable containment system. Conduit fittings such as these are readily available and a skilled electrician is able to present professional work using the equipment.

switch to switch and from light to light; joints should not be necessary and are to be discouraged.

The Use of Connectors

Small cables may be connected by means of a fixed connecting block with grub screws. Larger cables should be connected with substantial mechanical clamps (not grub screws) and the ends of the cables should preferably be fitted with cable lugs. Lug terminals must be large enough to contain all the strands of the conductor and should be connected together with bolts and nuts or bolted to connecting studs mounted on an insulated base. There are various other types of mechanical clamp available which are suitable for connecting large cables. The jointing of MI copper-sheathed cables requires special consideration, and the method is described in Chapter 14.

Crimping

It is common practice to use crimping lugs for terminating all sizes of copper and aluminium conductors. Crimping is carried out with special 'crimping lugs' and crimping tools. This makes a very efficient joint, and the need for special solders and fluxes is eliminated. Crimping has almost entirely replaced soldered joints.

However, to obtain satisfactory joints it is important to use the correct crimping tool and lug for the size of cable being jointed. The crimping tools

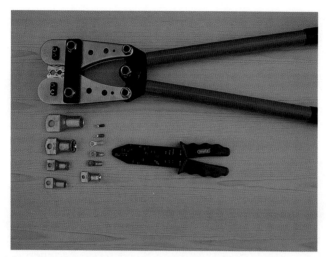

FIGURE 8.14 Cable crimping is commonly employed for cable terminations. The illustration shows small and large crimping termination lugs and some examples of crimping tool available (Highfield Engineering).

must also be kept in good condition and the dies inspected periodically, as any undue wear will result in unsatisfactory joints. This is particularly important where the connections will be subject to vibration, which can occur in a variety of industrial applications. British Standard BS 7609:1992 offers much useful

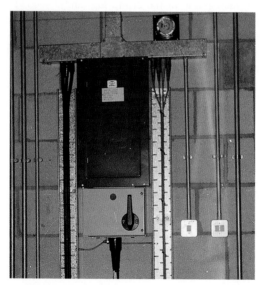

FIGURE 8.15 The illustration shows a number of wiring systems in use. Cables installed in conduit, trunking and on cable tray are present, and the bond between the cable tray to the trunking can also be seen. The fire alarm and fire alarm bell are wired in FP200 (fire performance) cable (William Steward & Co. Ltd).

FIGURE 8.16   The installation is completed, the competent electrician makes sure the site is left tidy. A vacuum cleaner is an essential part of the electrician's equipment (M.W. Cripwell Ltd).

advice on crimp and cable preparation, and the procedures to be used to obtain satisfactory joints.

There are two main types of crimping die, those which make a single depression in the facing side of the crimp lug, and a second type which, on compression, de-form the lug into a hexagonal shape. Either type of tool may be used for the majority of crimping purposes, but if using multi-stranded conductors such as those in 'tri-rated' cables, it is important to utilise the hexagonal form. This is so that correct compression takes place and an unsatisfactory joint is avoided.

Chapter 9

Conduit Systems

A conduit is defined as a tube or channel. In electrical installation work 'conduit' refers to metal or plastic tubing. The most common forms of conduit used for electrical installations are made to BS EN 61386, and these may be of steel or PVC plastic. Non-ferrous metallic conduits mainly in copper and aluminium were formerly used in special installations but have been virtually replaced by either the steel or plastic forms. Screwed steel conduit is a widely used and very effective cable containment system. Conduit fittings are readily available and a skilled electrician is able to present professional work using the equipment.

9.1 AN OVERVIEW OF CONDUIT INSTALLATION

The choice between steel or non-metallic conduit will be mainly influenced by site conditions, the use of the building, likely temperatures in the location and other factors such as likely exposure to corrosive or damp conditions. Screwed steel conduit offers good protection against mechanical damage and PVC materials are unaffected by moisture. Once the choice of material has been made, the next step is to select the most suitable 'runs' for the conduits. When there are several conduits running in parallel, they must be arranged to avoid crossing at points where they take different directions. The routes should be chosen so as to keep the conduits as straight as possible, only deviating if the fixings are not good. The 'runs' should also be kept away from gas and water pipes and obstructions which might prove difficult to negotiate. Locations where they might become exposed to dampness or other adverse conditions should be avoided.

Conduit Fittings

It is quite permissible to use manufactured bends, inspection tees and elbows but, for a neat appearance, there will be occasions where plain bends are better achieved by setting the conduit. Where there are several conduits running in parallel which change direction it is necessary for these bends to be made so that the conduits follow each other symmetrically. This is not possible if manufactured bends are used.

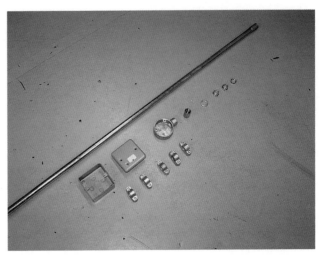

FIGURE 9.1   Components for a run of steel conduit, ready for installation (M.W. Cripwell Ltd).

At junctions, drawing in points and at accessory fixing positions, it is quite usual to use round boxes. These boxes have a tidy appearance, provide plenty of room for drawing in cables and can accommodate some slack cable which should be stowed in all draw-in points. For conduits up to 25mm diameter, the small circular boxes should be used. These have an inside diameter of 60mm. The larger circular conduit boxes are suitable for 32mm-diameter conduits.

Circular boxes are not suitable for conduits larger than 32mm, and for these larger sizes or where several conduit runs are terminated, rectangular boxes should be used. It would be impossible to draw large cables even into the larger type of circular box, as there would not be sufficient room to enable the final loop of the cable to be stowed into the box. Rectangular boxes vary in size and some types are far too short for easy drawing in of cables, and they should therefore be selected to suit the size of cables to be installed.

Where two or more conduits run in parallel, it is a good practice to provide at draw-in points an adaptable box which embraces all of the conduits. This presents a much better appearance than providing separate draw-in boxes and has the advantage of providing junctions in the conduit system which might prove useful if alterations have to be made at a later date. The conduit system for each circuit should be erected completely before any cables are drawn in.

An advantage of the conduit system is that the cables can be renewed or altered easily at any time. It is, therefore, necessary that all draw-in boxes should be readily accessible, and subsequently nothing should be fixed over or in front of them so as to render them inaccessible. The need for the conduit system to be complete for each circuit, before cables are drawn in, is to ensure

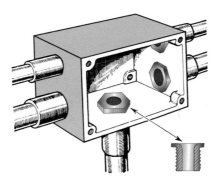

FIGURE 9.2   Where two or more conduits are run in parallel it is good practice to embrace all conduits with an adaptable box.

that subsequent wiring can be carried out just as readily as the original; it prevents cables becoming damaged where they protrude from sharp ends of conduit, and avoids the possibility of drawing the conduit over the cables during the course of erection.

The Radius of Conduit Bends

Facilities, such as draw-in boxes, must be provided so that cables are not drawn round more than two right-angle bends or their equivalent. The radius of bends must not be less than the standard normal bend for the size of conduit being used (Fig. 9.3).

Methods of Fixing Conduit

There are several methods of fixing conduit, and the one chosen generally depends upon what the conduit has to be fixed to.

Distance Saddles

Distance saddles are most commonly used and are fixed by means of screwing into the wall or other surface. They are designed to space conduits approximately 10mm from the wall or ceiling. Distance saddles are generally made of malleable cast iron. They are much more substantial than other types of saddles, and as they space the conduit from the fixing surface they provide better protection against corrosion.

When conduit is fixed to concrete, a high percentage of the installation time is spent in drilling and plugging for fixings. The use of distance or spacer bar saddle having only one fixing hole in its centre has an advantage over the ordinary saddle, in spite of the higher cost of the former.

The use of this type of saddle eliminates the possibility of dust and dirt collecting behind and near the top of the conduit where it is generally

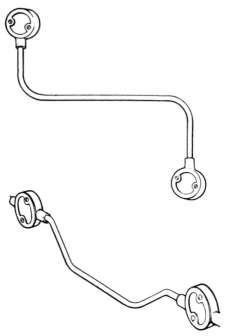

FIGURE 9.3 Cable must not be drawn round more than two right-angle bends or their equivalent. The four bends in the lower diagram are each at 45°, making a total of 180° in all.

inaccessible. Special 'hospital' saddles are obtainable and these increase the space between the conduit and the wall, greatly aiding cleaning. For this reason they are usually specified for hospitals, kitchens and other situations where dust traps must be avoided.

Spacer Bar Saddles

Spacer bar saddles are ordinary saddles mounted on a spacing plate. This spacing plate is approximately of the same thickness as the sockets and other conduit fittings and, therefore, serves to keep the conduit straight where it leaves these fittings. A function of the spacer bar saddles is to prevent the conduit from making contact with plaster and cement walls and ceilings which could result in corrosion of the conduit.

Some types of spacer bar saddles are provided with saddles having slots instead of holes. The idea is that the small fixing screws need only be loosened to enable the saddle to be removed, slipped over the conduit and replaced. This advantage is offset by the fact that when the saddle is fixed under tension there is a tendency for it to slip sideways clear of its fixing screws, and there is always a risk of this happening during the life of the installation if a screw should become slightly loose. For this reason holes rather than slots are generally more satisfactory in these saddles.

FIGURE 9.4   Saddles for use with steel conduit.

When selecting the larger sizes of spacer bar saddles it is important to make sure that the slotted hole which accommodates the countersunk fixing screw is properly proportioned. If they are not countersunk deep enough to enable the top of the screw to be flushed with the top of the spacing bar, an unsatisfactory conduit fixing may result.

Ordinary Saddles

Ordinary saddles are not extensively used. Fixing is by means of two screws. They provide a secure fixing and should be spaced not more than 1.3m apart. Conduit boxes to which luminaires are to be fixed should be drilled at the back and fixed, otherwise a saddle should be provided close to each side of the box. When ordinary saddles are used the conduit is slightly distorted when the saddle is tightened.

Multiple Saddles

Where two or more conduits follow the same route it is generally an advantage to use multiple saddles. The proper method is for the conduits to be spaced so that when they enter conduit fittings there is no need to set the conduit. An alternative means of running two or more conduits together is to stagger the saddle positions, allowing the conduits to be placed closer together.

Multiple spacer bar saddles can be purchased or they can be made up to suit a particular installation. Where several conduits have to be run on concrete, the use of multiple saddles saves a considerable amount of fixing time, as only two screws are required, and also ensures that all conduits are properly and evenly spaced (Fig. 9.5).

Girder Clips

Where conduits are run along or across girders, trusses or other steel framework, a number of methods of fixing may be used by the installer. A range

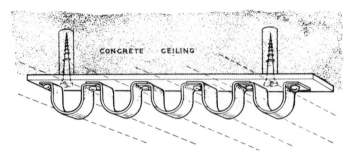

FIGURE 9.5 A 25mm × 5mm steel strip here is used to support five conduits on a concrete ceiling. It has two screw fixings.

of standard spring clips is available which can be quickly and easily fitted (Fig. 9.6).

Other methods are also available including a range of bolt-on devices. If it is intended to run a number of conduits on a particular route and standard clips are not suitable, it may be advisable to make these to suit site conditions. Multiple girder clips can be made to take a number of conduits run in parallel (Fig. 9.7).

Under certain circumstances, as an alternative to girder clips, conduit fixings can be welded to steelwork, or the steelwork could be drilled. However, structural steelwork should never be drilled or welded unless approval for this is obtained from the structural engineer.

When conduits are suspended across trusses or steelwork there is a possibility of sagging, especially if luminaires are suspended from the conduit

FIGURE 9.6 A selection of standard clips designed for quick fitting of conduit to girders and other steelwork.

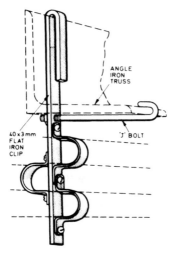

FIGURE 9.7 Supporting several conduits from angle iron truss.

between the trusses. These conduits should either be of sufficient size to prevent sagging, or be supported between the trusses. They can sometimes be supported by steel rods from the roof above. If the trusses are spaced 3m or more apart it is not very satisfactory to attempt to run any conduit across them, unless there is additional means of support. It is far better to take the extra trouble and run the conduit at roof level where a firm fixing may be found.

TABLE 9.1 Spacing of Supports for Conduits (Extract from IEE On-Site Guide, Table 4C)

Nominal size of conduit (mm)	Maximum distance between supports					
	Rigid metal		Rigid insulating		Pliable	
	Horizontal (m)	Vertical (m)	Horizontal (m)	Vertical (m)	Horizontal (m)	Vertical (m)
Not exceeding 16	0.75	1.0	0.75	1.0	0.3	0.5
Exceeding 16 and not exceeding 25	1.75	2.0	1.5	1.75	0.4	0.6
Exceeding 25 and not exceeding 40	2.0	2.25	1.75	2.0	0.6	0.8
Exceeding 40	2.25	2.5	2.0	2.0	0.8	1.0

Conduit Cutting

Conduit should be cut with a hacksaw in preference to a pipe cutter, as the latter tends to cause a burr inside the conduit. In any case, the ends of all conduits must be carefully reamered inside the bore with a file, or reamer, to be certain that no sharp edges are left which might cause damage to the cables when they are being drawn in. This reamering should be carried out after the threading has been completed (Fig. 9.8).

Drilling and Cutting

Other useful tools include electric drills, for drilling fixing holes and also for drilling holes in conduit fittings and distribution boards. Suitable hole cutters for cutting holes can be used in electric drills. Electric hammer-drills save a considerable amount of time for wall plugging, cutting holes through walls and floors. Safety eye shields must be worn. As a final word of advice, do not forget to make good the holes after the conduit has been erected.

Checking Conduit for Obstructions

When the length of conduit has been removed from the pipe vice, it is a good idea to look through the bore to ensure that there are no obstructions. Some foreign object, such as a stone, may have entered the conduit during storage (especially if stored on end) or welding metal may, in rare cases, have become deposited inside the conduit. If such obstructions are not detected before the installation of the conduit, considerable trouble may be experienced when the cables are drawn in.

FIGURE 9.8 Conduit should be cut using a hacksaw and after cutting and threading any burrs should be removed using a file.

Running Conduit in Wooden Floors

Where conduit is run across the joists, they will have to be slotted to enable the conduit to be kept below the level of the floorboards. When slots are cut in wooden joists they must be kept within 0.1–0.2 times the span of the joist, and the slots should not be deeper than 0.15 times the joist depth. The slots should be arranged so as to be in the centre of any floorboards, if they are near the edge there is the possibility of nails being driven through the conduit. Floor 'Traps' should be left at the position of all junction boxes. These traps should consist of a short length of floorboard, screwed down and suitably marked.

Running Conduit in Solid Floors or Ceilings

Where there are solid floors the conduit needs to be arranged so that cables can be drawn in through ceiling or wall points. This method is known as the 'looping-in system', and it is shown in Fig. 9.10. Conduit boxes are provided

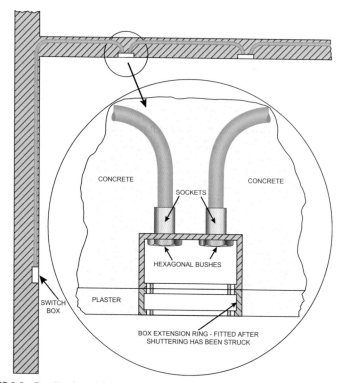

FIGURE 9.9   Details of conduit box and method of fastening conduit. A socket is fixed outside the back of the box and a brass hexagonal bush inside. The bush should be firmly tightened, otherwise there will be difficulty in obtaining a satisfactory continuity test on completion.

with holes at the back to enable the conduit to be looped from one box to another. These boxes are made with two, three or four holes so that it is possible also to tee off to switches and adjacent ceiling or wall points.

The correct method of fixing the conduit to these boxes is to fit a socket outside the back of the box, and a hexagon brass bush inside. The bush should be firmly tightened with a special box spanner, and consultants very often specify that lead, neoprene or copper compression washers are fitted between the bush and the box. If these joints in the conduit system are not absolutely tight there might be difficulty in obtaining a satisfactory continuity test on completion. Satisfactory continuity is essential.

There are many types of solid floor and some of these are so shallow that a very sharp set has to be made in the conduit after it leaves the socket. For this it may be necessary to use a bending machine with a special small former made for these types of floors.

If the floors are of reinforced concrete, it may be necessary to erect the conduit system on the shuttering and to secure it in position before the concrete is poured. If not securely fixed, it may move out of position or lift and then, when the concrete is set, it will be too late to rectify matters. Wherever conduit is to be buried by concrete, special care must be taken to ensure that all joints are tight and secure, otherwise liquid cement may enter the conduit and form a solid block inside. Joints should be painted with bitumastic paint, and the conduit itself should also be painted where the enamel has been removed during threading or setting. In the case of galvanised conduit, the paint should be a zinc rich cold galvanising coating such as 'Galvafroid'.

Sometimes the conduits can be run in chases cut into concrete floors; these should be arranged so as to avoid traps in the conduit where condensation may collect and damage the cables.

Conduit Runs to Outlets in Walls

Sockets near skirting level should preferably be fed from the floor above rather than the floor below, because in the latter case it would be difficult to avoid traps in the conduit (Fig. 9.10).

When the conduit is run to switch and other positions in walls it is usually run in a chase in the wall. These chases must be deep enough to allow at least 10mm of cement and plaster covering. Steel conduits buried in plaster should be given a coat of protective paint, or should be galvanised if the extra cost is justified.

Make sure that the plaster is finished neatly round the outside edges of flush switch and socket boxes, otherwise the cover plates may not conceal any deficiencies in the plaster finish. When installing flush boxes before plastering, it is advisable to stuff the boxes with paper to prevent their being filled with plaster.

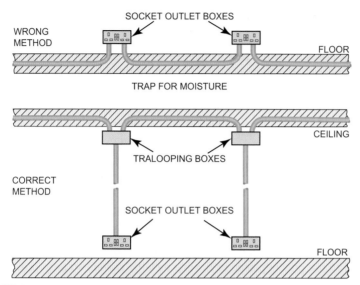

FIGURE 9.10 Right and wrong methods of feeding sockets near skirting level. If the sockets are fed from the floor below, it is difficult to avoid a trap for moisture.

FIGURE 9.11 A metal conduit box with an extension piece for use where the depth has to be increased as building work progresses. See Fig. 9.9.

Ceiling Points

At ceiling points the conduit boxes will be flushed with the finish of the concrete ceiling. If the ceiling is to have a plaster rendering, this will leave the front of the boxes recessed above the plaster finish. To overcome this it is possible to purchase extension rings for standard conduit boxes.

At the position of ceiling points it is usual to provide a standard round conduit box, with an earth terminal, but any metal box or incombustible enclosure may be used, although an earth terminal must be provided.

Running Sunk Conduits to Surface Distribution Boards

Where surface mounted distribution boards are used with a sunk conduit, the problem arises as to the best method of terminating flush conduits into the surface boards. The best method is to fit a flush adaptable box in the wall behind the distribution board, and to take the flush conduits directly into it. Holes can be drilled in the back of the distribution board and bushed. Spare holes should be provided for future conduits. Alternatively, an adaptable box can be fitted at the top of the distribution board, partly sunk into the wall to receive the flush conduits, and partly on the surface to bolt on the top of the distribution board. Distribution boards must be bonded to the adaptable boxes.

Flexible Conduit

For final connections to motors, or any similar equipment liable to vibration, it is usual to use pliable plastic or flexible metallic conduit so as to provide for movement. It also prevents any noise or vibration being transmitted from the motor, or the machine to which it may be coupled, to other parts of the building through the conduit system.

This flexible conduit should preferably be of the watertight pattern and should be connected to the conduit by means of suitable adaptors. These adaptors are made to screw on to the conduit to secure the flexible tubing. A sound connection is essential, as otherwise it is likely to become detached and expose the cables to mechanical damage.

The use of flexible metallic tubing which is covered with a PVC sleeving is recommended, as this outer protection prevents oil from causing damage to the rubber insertion in the joints of the tubing. An alternative is to use pliable plastic conduit, and, in either case, an appropriate CPC must be provided.

Conduit Capacity

The number of cables drawn into a particular size conduit should be such that no damage is caused to either the cables or to their enclosure during installation. It will be necessary, after deciding the number and size of cables to be placed in a particular conduit run, to determine the size of conduit to be used. Each cable and conduit size is allocated a factor and by summing the factors for all the cables to be run in a conduit route, it is an easy matter to look up the appropriate conduit size to use.

For example, if it is desired to run eight 2.5mm^2 and four 4.0mm^2 cables along a 4m run of conduit with two bends, it is possible to determine the conduit size as follows.

From the IEE On-Site Guide, Table 5C, factors for 2.5mm^2 and 4.0mm^2 cables are 43 and 58, respectively.

$$8 \times 43 = 344$$
$$4 \times 58 = 232$$
$$\overline{\text{Total} = 576}$$

From the IEE On-Site Guide, refer to Table 5D, for a 4m run with two bends. As can be seen 25mm-diameter conduit with a factor of 388 would be too small; 32mm-diameter conduit with a factor of 692 will be suitable.

It must always be remembered that, as the number of cables or circuits in a given conduit or trunking increases, the current-carrying capacities of the cables decrease. It may therefore be advisable not to increase the size of the conduit in order to accommodate more cables, but to use two or more conduits.

9.2 THE SCREWED STEEL CONDUIT SYSTEM

The foregoing sections relating to conduit installation apply to both steel and PVC types. However, some additional notes are warranted and this section deals specifically with the screwed steel conduit system. It is commonly used for permanent wiring installations, especially for commercial and industrial buildings. Its advantages are that it affords the conductors good mechanical protection, permits easy rewiring when necessary, minimises fire risks, and presents a pleasing appearance if properly installed. Correct installation is important, and the general appearance of a conduit system reflects the degree of skill of the person who erected it.

The disadvantages are that it is expensive compared with some systems, is difficult to install under wood floors in houses and flats and is liable to corrosion when subjected to acid, alkali and other fumes. Moreover, under certain conditions, moisture due to condensation may form inside the conduit.

Protection of Conduit

Although heavy gauge conduit affords excellent mechanical protection to the cables it encloses, it is possible for the conduit itself to become damaged if struck by heavy objects. Such damage is liable to occur in workshops where the

TABLE 9.2 Steel Conduit Dimensions

Nearest Imperial size (in)	Metric size (mm)	Thickness of wall	Pitch of thread (mm)
$\frac{5}{8}$	16	1.6	1.5
$\frac{3}{4}$	20	1.6	1.5
1	25	1.8	1.5
1¼	32	1.8	1.5

conduit is fixed near the floor level and may be struck by trolleys or heavy equipment being moved or slung into position. Protection can be afforded by threading a water pipe over the conduit during erection, or by screening it with sheet steel or channel iron. Another method of protection is, of course, to fix the conduit behind the surface of the wall.

Conduit Installed in Damp Conditions

If steel conduits are installed externally, or in damp situations, they should be galvanised and all clips and fixings (including fixing screws) shall be of corrosion-resisting material [IEE Regulation 522.3.1]. In these situations precautions must be taken to prevent moisture forming inside the conduit due to condensation. This is most likely to occur if the conduit passes from the outside to the inside of a building, or where there is a variation of temperature along the conduit route. In all positions where moisture may collect, holes should be drilled at the lowest point to allow any moisture to drain away. Drainage outlets should be provided where condensed water might otherwise collect. Whenever possible conduit runs should be designed so as to avoid traps for moisture.

Continuity of the Conduit System

A screwed conduit system must be mechanically and electrically continuous across all joints so that the electrical resistance of the conduit, together with the resistance of the earthing lead, measured from the earth electrode to any position in the conduit system shall be sufficiently low so that the earth fault current operates the protective device. To achieve this it is necessary to ensure that all conduit connections are tight, and that the enamel is removed from adaptable boxes and other conduit fittings where screwed entries are not provided. To guarantee the continuity of the protective conductor throughout the life of the installation, it is common practice to draw a separate circuit protective conductor into the conduit for each circuit in the conduit.

Some Practical Hints

Apart from the electrician's ordinary tools, such as rule, hacksaw, hammer, screwdriver, pliers etc., it is necessary to have stocks and dies, file or reamer, bending machine and a pipe vice. For 16mm and 20mm conduit, the small stocks are suitable, but for 25mm and 32mm, the medium stocks should be used. Although 25mm dies are provided for the small stocks it is best to use the medium stocks for 25mm conduit.

Stocks and dies for screwing conduit should be clean, sharp and well lubricated, and should be rotated with a firm and steady movement. To get the best results stocks and dies should be of the self-clearing pattern to prevent the soft swarf from clogging the chasers. Worn dies and guides should always be replaced when showing signs of wear, otherwise the workmanship will suffer as a result of bad threads.

Ratchet operated stocks and dies are available which are useful for the larger thread sizes and there are also powered conduit screwing machines which offer certain advantages on a conduit installation where a considerable amount of large conduit is being installed.

Bending Conduit

It is normal to use a bending machine for all sizes of conduit which enable bends and sets to be made without the risk of kinks or flattening of the conduit. These machines are also necessary when very sharp bends have to be made in 16mm and 20mm conduits.

Avoidance of Gas, Water and Other Pipes

All conduits must be kept clear of gas and water pipes, either by spacing or insulation. Although, conduits may make contact with water pipes provided that they are intentionally bonded to them. They must not make casual contact with water pipes. The reason for this precaution is that if the conduit system reaches a high potential due to defective cables in the conduit and an ineffective earth continuity, and this conduit makes casual contact with a gas or water pipe, either of which would be at earth potential, then arcing would take place between the conduit and the other pipe. This might result in puncturing the gas pipe and igniting the gas.

9.3 SCREWED COPPER CONDUIT

Sometimes, for very high integrity installations, copper conduit is used. The advantage of copper conduit is that it resists corrosion and provides excellent continuity, but for normal installation work, the cost could prove to be prohibitive.

Copper conduit can be screwed in the same manner as steel conduit although the screwing of copper is more difficult than mild steel. Connections are generally made by soldering and bronze junction boxes should preferably be used.

This system is comparatively expensive, but is used in buildings where long life and freedom from corrosion of the conduit and the cables are of first importance. For example, some state buildings are provided with copper conduits where the conduit system is buried in concrete floors and walls.

9.4 INSULATED CONDUIT SYSTEM

Non-metallic conduits are being increasingly used for all types of installation work, both for commercial and domestic wirings. The PVC rigid conduit is made in all sizes from 16mm to 50mm in external diameter, and there are

(a) The conduit drop in this example is being run from overhead trunking and is to be used to feed a new hand dryer. The first step is to measure the distance from the outlet box to the wall to determine the position of the bend required. (b) Having marked the position required for the bend, the conduit is placed in the bending machine.

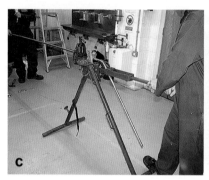

(c) Using an assistant to steady the conduit, the tube is bent to the required angle. (d) The position of the conduit is marked using a plumb line and the position of the saddles is marked. The wall is drilled for the saddle fixings.

(e) Suitable saddles are screwed into position. (f) After determining the position of the outlet box, the conduit is cut to length using a hacksaw. The conduit bending machine incorporates a conduit vice which is useful for securing the conduit during cutting, threading and assembly of fittings.

FIGURE 9.12 Installing a steel conduit drop.

(g) The conduit is temporarily clipped in position and the trunking marked and drilled for the outlet box fixing. In this case there are already cables in the trunking and these are moved to a safe position and secured before drilling. (h) Sharp edges caused by drilling are removed using a file.

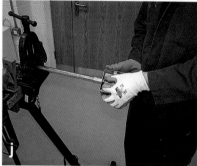

(i) Using a set of dies and a die holder, the conduit is threaded ready of the outlet bush. Application of a thread cutting lubricant aids this process. (j) After the coupler is secured, the outlet box is attached using a brass bush.

(k) After fixing the conduit and screwing the box to the wall, the bush is tightened with a bush spanner. (l) To secure the bush fully, a bush spanner should be used. These are available in a range of sizes (all M.W. Cripwell Ltd).

FIGURE 9.12   cont'd.   Installing a steel conduit drop.

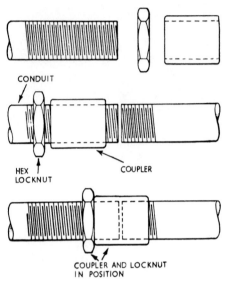

FIGURE 9.13   For connecting two lengths of conduit, neither of which can be turned. The method of using the coupler and locknut may be clearly followed.

various types of conduit fittings, including boxes available for use with this conduit.

Figure 9.14 shows a range of typical components made of a plastic material, and fitted with special sockets which enable the conduit to be assembled neatly. Solvent adhesives are available for jointing though a suitable number of sliding joints must be left to allow for expansion.

The advantage of the insulated conduit system is that it can be installed much more quickly than steel conduit, it is non-corrosive, impervious to most chemicals, weatherproof, and it will not support combustion. The disadvantages are that it is not suitable for temperatures below −5 °C, or above 60 °C, and where luminaires are suspended from PVC conduit boxes, precautions must be taken to ensure that the heat from the lamp does not result in the PVC box reaching a temperature exceeding 60 °C.

For surface installations it is recommended that saddles be fitted at intervals of 800mm for 16mm-diameter conduit, and intervals of 1600–2000mm for larger sizes. The special sockets and saddles for this type of conduit must have provision to allow for longitudinal expansion which may take place with variations in ambient temperature.

It is of course necessary to provide a circuit protective conductor in all insulated conduits, and this must be connected to the earth terminal in all boxes for switches, sockets and luminaires. The only exception is in connection with Class 2 equipment, i.e. equipment having double insulation. In this case

FIGURE 9.14 A range of fittings for use with plastic conduit includes the boxes, couplers, saddles and clips (M.W. Cripwell Ltd).

a protective conductor must not be provided except as covered in IEE Regulation 412.2.2.4.

Flexible PVC conduits are also available, and these can be used with advantage where there are awkward bends, or under floorboards where rigid conduits would be difficult to install.

Installation of Plastic Conduit

Plastic conduits and fittings can be obtained from a number of different manufacturers and the techniques needed to install these are not difficult to apply. Care is, however, needed to assemble a neat installation and the points given below should be borne in mind. As with any other installation good workmanship and the use of good quality materials are essential.

It should be noted that the thermal expansion of plastic conduit is about six times that of steel, and so whenever surface installation of straight runs exceeding 6m is to be employed, some arrangement must be made for expansion. The saddles used have clearance to allow the conduit to expand. Joints should be made with an expansion coupler which is attached with solvent cement to one of the lengths of tube, but allowed to move in the other.

Cutting the conduit can be carried out with a fine tooth saw or using a special tool designed for the purpose. As with steel conduit, it is necessary to remove any burrs and roughness at the end of the cut length.

Bending the small sizes of plastic conduit up to 25mm diameter can be carried out cold. A bending spring is inserted so as to retain the cross-sectional shape of the tube. It is important to use the correct size of bending spring for the type of tube being employed. With cold bending, the tube should initially be bent to about double the required angle, and then returned to the angle required, as this reduces the tendency of the tube to return to its straight form. To bend larger sizes of tube, 32mm diameter and above, judicious application of heat is needed. The formed tube should as soon as possible be saddled after bending.

Joints are made using solvent adhesives which can be obtained specifically for the purpose. These adhesives are usually highly flammable and care is needed in handling and use. Good ventilation is essential, and it is important not to inhale any fumes given off. Clean and dry the components to be joined before

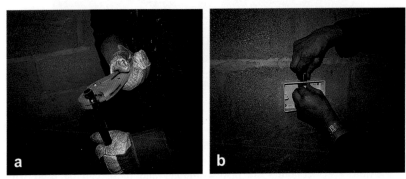

(a) The conduit can easily be cut to length with the special cutters obtainable for the purpose. (b) A range of fittings is available including boxes, bends, saddles and bushes. Here a bush is being secured using a threaded ring.

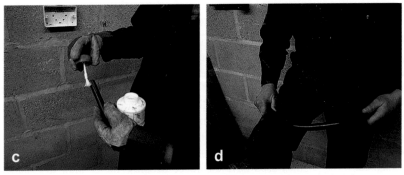

(c) Jointing is carried out by the use of solvent adhesives, and care must be taken in their application to avoid blocking small-sized conduits. (d) Bending can be by hand and a bending spring is used to retain the shape of the bore. In cold conditions, the conduit should be warmed before being bent through an angle slightly greater than that required. It is then returned to the right angle generally needed for bends.

FIGURE 9.15   Stages in the assembly of plastic conduit.

Continued

(e) Fixing of boxes and saddles is carried out using wall plugs and screws, as with steel conduit.
(f) Pre-formed bends are available and these are fitted with clip on lids, as shown here
(all Highfield Engineering).

FIGURE 9.15   cont'd.   Stages in the assembly of plastic conduit.

commencing work. Avoid using excess solvent as this may block the conduit by
forming a barrier across the inside especially when joining small size conduits.
Using too little solvent may not make a waterproof joint. Experience will
indicate the correct quantity of adhesive to use. The manufacturers' instructions
for use of the solvent adhesive should be strictly followed. It is generally
necessary to apply the adhesive to both surfaces to be joined, pushing the
components together and holding them steady for about 15s without moving to
ensure the joint is set. Where expansion joints are needed the expansion collar
should be solvent welded to one length of tube, but left free to slide on the other.
If sealing is needed to waterproof the joint, use a special non-setting adhesive or
grease. Threaded adaptors are available for use when it is required to make
connections to screwed systems. These can be solvent welded to the plastic tube
and screwed into the threaded fitting as required.

The use of plastic conduit is suitable when cable runs require to be located in
pre-cast concrete. As will be appreciated it is essential that sound joints are
made so that when the concrete is cast, the conduit runs do not separate.

9.5 CABLES IN CONDUITS

The types of cables which may be installed in conduits are PVC single-core
insulated, butyl or silicone rubber insulated, with copper or aluminium
conductors. PVC insulated and sheathed cables are sometimes installed in
conduits when the extra insulation and protection are desirable or sometimes,
because it is simply more convenient.

Typically, cables are rated at 600/1000V. The metric cables are smaller in
overall diameter than the Imperial sizes, due to a reduction in the thickness of
the insulation, and, a larger number of cables may be drawn into a given size of
conduit, than was permissible for the Imperial sizes.

One word of warning is necessary; cables up to 2.5mm^2 may have solid conductors, and it has been found that these are not so easily drawn into conduit as the earlier type of stranded cables.

This does not apply to butyl-insulated cables which will be supplied as seven strand conductors for 1.5mm^2 and 2.5mm^2 cables. It is, however, possible to install eight 2.5mm^2 cables in 20mm conduit and therefore two ring circuits can be accommodated. Sixteen-millimetre conduit will accommodate six 1.5mm^2 PVC insulated cables, but this size of conduit is rarely used in practice.

Removal of Burrs from Ends of Conduit

As described in Section 3, removal of burrs from the ends of cut conduit is essential to prevent damage to cables. This must be done after the conduit is cut and screwed and before it is assembled.

Drawing Cables into Conduits

Cables must not be drawn into conduits until the conduit system for the circuit concerned is complete, except for prefabricated 'modular' flexible conduit systems which are not wired *in situ*.

When drawing in cables they must first of all be run off the reels or drums, or the reels must be arranged to revolve freely, otherwise if the cables are allowed to spiral off the reels they will become twisted, and this would cause damage to the insulation. If only a limited quantity of cable is to be used it may be more convenient to dispense it direct from one of the boxed reels which are on the market. If a number of cables are being drawn into conduit at the same time, the cable reels should be arranged on a stand or supported in a vice so as to allow them to revolve freely (Figs 9.16 and 9.17).

In new buildings and in damp situations the cable should not be drawn into conduits until it has been made certain that the interiors of the conduits are dry and free from moisture. If in doubt, a draw wire with a swab at the end should be drawn through the conduit so as to remove any moisture that may have accumulated due to exposure or building operations.

Unless the runs are quite short, it is usual to commence drawing in cables from a mid-point in the conduit system so as to minimise the length of cable which has to be drawn in. A draw-in tape should be used from one draw-in point to another and the ends of the cables attached. The ends of the cables must be bared for a distance of approximately 50mm and threaded through a loop in the drawtape. When drawing in a number of cables they must be fed in very carefully at the delivery end whilst someone pulls them at the receiving end.

The cables should be fed into the conduit in such a manner as to prevent any cables crossing, and also to avoid them being pulled against the sides of the opening of the draw-in box. In hot weather or under hot conditions, the drawing-in can be assisted by using cable-pulling lubricant. Always leave

FIGURE 9.16 Running off cables from reels. Illustrates a typical method used when only a few cables are involved. A short piece of conduit is gripped in a pipe vice. When many reels have to be handled it is best to use a special rack.

some slack cable in all draw-in boxes and make sure that cables are fed into the conduit so as not to finish up with twisted cable at the draw-in point.

This operation needs care and there must be synchronisation between the person who is feeding and the person who is pulling. If in sight of each other this can be achieved by a movement of the head, and if within speaking distance by word of command given by the person feeding the cables. If the two persons

FIGURE 9.17 Cable must not be allowed to spiral off reels or it will become twisted and the insulation damaged. It should be run off by a method similar to that shown in Fig. 9.16.

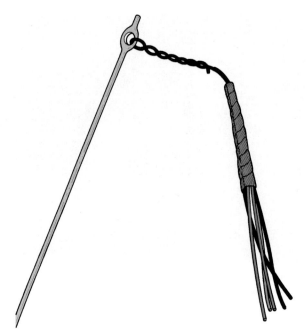

FIGURE 9.18 How to connect cable to draw tape.

are not within earshot, then the process is somewhat more difficult. A good plan is for the individual feeding the cables to give pre-arranged signals by tapping the conduit with a pair of pliers or similar metallic object. In some cases, it may be necessary for a third person to be stationed midway between the two positions to relay the necessary instructions from the person feeding to the person pulling. If cables are not drawn in carefully in this manner, they will almost certainly become crossed and this might result in the cables becoming jammed inside the conduit. In any case, it would prevent one or more cables being drawn out of the conduit should this become necessary (Fig. 9.20).

Looping in

When wiring an installation with PVC covered cable in conduit, joints are avoided as far as possible, and the looping-in system is normally adopted. In practice when wiring in conduit, the two lengths of cable forming the loop are threaded in separately and the junction is made at the switch, light or other terminal.

Before Wiring Sunken Conduit

Before wiring, the conduits for each circuit must be erected complete. Not only should they be complete but they must be clean and dry inside otherwise the

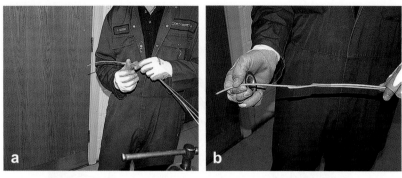

(a) The cables are staggered and taped in such a way that each will enter the conduit in turn. (b) The taped cable cores ready for attachment to the draw tape.

(c) The draw tape is fed into the conduit ready to pull in the cables. (d) With the help of an assistant, the cables are attached to the draw tape.

(e) Feeding should be done in such a manner as to prevent any cables crossing or becoming twisted. The operation needs care, and there must be synchronisation between the person feeding and the person who is pulling. The drawing-in can be assisted by using a cable-pulling lubricant. (f) After leaving some slack at the draw-in box, the cables can be cut to length (all M.W. Cripwell Ltd).

FIGURE 9.19 Drawing cables into conduit.

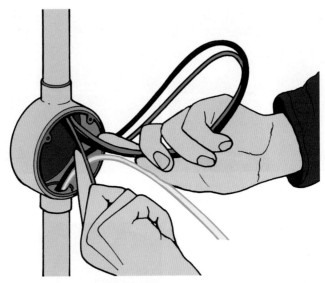

FIGURE 9.20 When a large number of cables have to be threaded at the same time, two hands are needed. The illustration shows the method of gripping the cable so as to guide into the conduit with the two forefingers.

FIGURE 9.21 A neat completed conduit installation, ready for wiring to be installed (William Steward & Co. Ltd).

cables may suffer damage. No attempt should be made to wire conduits which are buried in cement until the building has dried out and then the conduits should be swabbed to remove any moisture or obstructions which may have entered them.

Chapter 10

Trunking Systems

10.1 AN OVERVIEW OF TRUNKING INSTALLATION

A number of reputable manufacturers can supply trunking ranging from 25mm × 25mm to 300mm × 300mm or even larger, with two or more compartments. They also provide all necessary accessories such as bends, tees, crossovers and bridges to segregate cables of different systems at junctions.

Trunking systems are more flexible than conduit systems. Extensions can readily be made during the life of the installation by making a new hole in the trunking and running a conduit to a new point. Naturally, care is needed with the design of such an alteration as grouping of additional circuits may require the de-rating of cables to be re-assessed. However, it may be possible to implement the alteration without disturbing the existing wiring.

Trunking can be easily and quickly erected, and can be fitted to walls or suspended across trusses; where it should be supported at each joint. As with conduit, guidance on the spacing of supports for conduit is given in the IEE On-site Guide. IEE Table 4D covers both steel and plastic trunking types.

Where there are vertical runs of trunking, pin racks should be fitted inside the trunking to support the weight of the cables and to enable the cables to be secured during installation. These pin racks consist of steel pins, sheathed by an insulating material, mounted on a backplate; they should be fitted at intervals of 5m.

Where vertical trunking passes through floors it must be provided with internal fire barriers, which must consist of non-flammable materials, cut away to enable cables to pass through and made good after the installation of the cables.

When large cables are installed in trunking, care must be taken to ensure that all bends are of sufficient radius to avoid damaging the cable (IEE Regulation 522.8.3). The IEE On-site Guide gives useful advice on this subject. This states, for example, that non-armoured PVC-insulated cables of an overall diameter greater than 25mm shall not be so bent that the radius of the inside of the bend is less than six times the diameter of the cable. Trunking manufacturers provide bends and tees that enable this requirement to be satisfied.

Trunking can be used to accommodate PVC insulated cables that are too large to be drawn into conduit. Unless there are special reasons for using conduit, it will generally be found more economical to use trunking rather than conduit larger than 32mm diameter.

Regulations

The regulations governing wiring in conduit also apply to wiring in trunking, as far as applicable. All sections of trunking, bends, and other accessories must be effectively earthed in order to ensure that the conductivity of the trunking is such as to enable earth-fault current to flow to operate the fuse or earth-leakage circuit breaker protecting the circuit.

Trunking is usually supplied in 3 m lengths, although in some cases longer lengths are obtainable. If copper links are fitted these will generally ensure satisfactory earth continuity, but if tests prove otherwise an insulated protective conductor should be installed inside the trunking. It is in any case common practice to provide separate circuit protective conductors to ensure earth continuity throughout the life of the installation. As with conduit the cable capacities of trunking can be calculated. To ensure that cables can be readily installed, a space factor of 45% should be used.

When a large number of cables are installed in trunking, due regard must be paid to temperature rise due to grouping of cables. IEE Tables 4C1–4C5 give details of the factors to be taken into account when cables are bunched in trunking or conduits, and in some circumstances this could result in a very considerable reduction in the current ratings of the cables installed in the trunking.

For example, if eight circuits are enclosed in trunking the correction factor, according to IEE Table 4C1, could be as much as 0.52 to the rating values for 16 single-core cables.

The ratings of cables installed in trunking are also affected by ambient temperatures, and a de-rating of PVC insulated cables will be necessary if the ambient temperature exceeds 30 °C, as will be seen by referring to the rating factors in IEE Tables 4B1–4B3.

Details of the application of correction factors for grouping and ambient temperature are given in Chapter 4.

10.2 METALLIC TRUNKING

Metallic trunking for industrial and commercial installations is often used in place of the larger sizes of conduit. It can be used with advantage in conjunction with 16mm–32mm conduits, the trunking forming the background of frame-work of the system with conduits running from the trunking to lighting or socket outlet points. For example, in a large office building, trunking can be run above the suspended ceiling along the corridors to feed corridor points, and rooms on either side can be fed from this trunking by conduits.

In multi-storey buildings trunking of suitable capacity, and with the necessary number of compartments, can be provided and run vertically in the riser ducts and connected to distribution boards; it can also accommodate circuit wiring, control wiring, also cables feeding fire alarms, telephones, emergency lighting and other services associated with buildings.

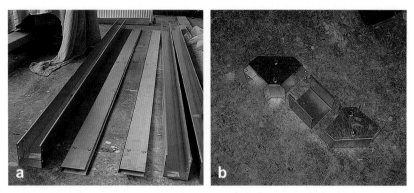

(a) 100 × 100mm steel trunking and lids laid out ready for erection. (b) Components for vertical and horizontal bends are available pre-formed and ready for assembly. The short straight length is cut to suit and is drilled with clearance holes for the fixing screws.

(c) Fitting the components of the offset together using set screws inserted into the pre-tapped holes in the angle pieces. (d) Having fitted the long runs of 100×100mm conduit to the wall, the offset is assembled in position.

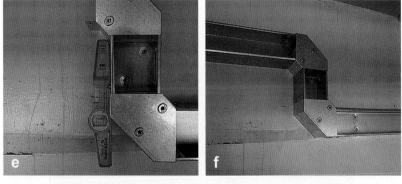

(e) Using a magnetic spirit level to check the alignment of the vertical section before tightening the fixing screws. Note the washers used under the heads of the fixing screws to spread the load. (f) The completed trunking assembly, with the offset, ready for wiring. Following which the lids will be fitted.

FIGURE 10.1 Steel trunking.

As explained in Chapter 8, cables feeding fire alarms and emergency circuits need to be segregated by fire-resisting barriers from those feeding low-voltage circuits (i.e. 50V–1000V). It is usual for telecommunications companies to insist that their cables are completely segregated from all other wiring systems. It may therefore be necessary to install three- or four-compartment trunking to ensure that IEE Regulation 528.1 and the requirements for data and telecommunications circuits are complied with. Cables feeding emergency lighting and fire alarms must also be segregated so as to comply with the requirements of BS 5266 and 5389. Additionally, segregation may be required to achieve electromagnetic compatibility requirements.

Lighting Trunking System

Steel or alloy lighting trunking was originally designed to span trusses or other supports in order to provide an easy and economical method of supporting luminaires in industrial premises at high levels.

The first types of such trunking consisted of extruded aluminium alloys, the sections of which were designed to support the weight of luminaires between spans of up to 5m. More recently sheet-steel trunking has become available, made in sections which achieve the same purpose.

The advantage of this type of trunking is that it can be very easily installed across trusses, will accommodate all wiring to feed the lighting points, and can also accommodate power wiring and, if fitted with more than one compartment, fire alarm and extra low-voltage circuits.

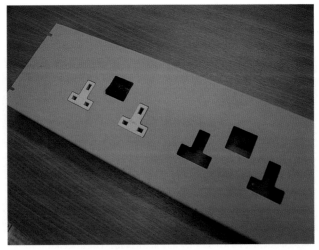

FIGURE 10.2 Shallow sockets can be obtained for fitting to the lid of skirting trunking and trunking manufacturers will punch suitable apertures for the reception of sockets.

When installed at high levels it can be very usefully employed to accommodate wiring for high-level unit heaters, roof fans and similar equipment. Its main purpose of course is to support luminaires, and when suspended between trusses, which have a maximum spacing of 5 m, it should be able to support the weight of the required number of luminaires without intermediate supports.

It is therefore necessary that trunking suspended in this manner is of sufficient size to take the necessary weight without undue deflection. Manufacturers of trunking provide the relevant data and should be consulted about this.

Lighting trunking is also manufactured in lighter and smaller sections which can be fixed directly to soffits, either on the surface or mounted flush with the finished ceiling; as this does not have to support heavy weights between spans it is similar to ordinary cable trunking.

Like all other trunking, it must be provided with suitable copper links between sections to ensure adequate earth continuity, but as already explained, if the earth continuity is found to be unsatisfactory, an insulated protective conductor should be installed in the trunking.

Some types of lighting trunking are of sufficient dimensions to accommodate the fluorescent lamps and control gear within the trunking. Others have the control gear in the trunking and the lamp fittings fixed beneath.

Steel Floor Trunking

Underfloor trunking made of steel is used extensively in commercial and similar buildings, and it can be obtained in very shallow sections with depth of only 22mm, which is very useful where the thickness of the floor screed is limited.

FIGURE 10.3 Office lighting fitted in integral trunking which houses the control units as well as the light fittings, the whole being suspended from the roof structure.

It is supplied with one or more compartments, and with junction boxes that have cover plates fitted flush with the level of the finished floor surface. Where there are two or more compartments these boxes are fitted with flyovers to enable Band I and Band II circuits to be kept segregated as required by IEE Regulation 528.1.

When floor ducts are covered by floor screed it is necessary to ensure that there is a sufficient thickness of screed above the top of the ducts to prevent the screed cracking as a result of the expected traffic on the floors. Another method is to use floor trunking, the top cover of which is fitted flush with the finished floor surface. In this case the top cover plate has to be of sufficient thickness to form a load-bearing surface.

Outlets for sockets and other points can be fitted on top of the cover plates, and it is usual to fit pedestals to accommodate the sockets.

Trunking is available which has sufficient depth to accommodate the socket and plugs, together with the necessary wiring. The minimum depth for this type of trunking is 50mm. Separate short sections of cover plate are provided in all positions where sockets may be required; these sections are easily removable and are provided with bushed holes to enable flexible cords to emerge. It is necessary to provide suitable holes in linoleum or carpets for the flexible cords to pass through.

Whatever type of floor trunking is employed, it can be connected to distribution board positions, and also to skirting trunking. Special right-angle bends are available to facilitate connection between floor trunking and skirting trunking.

If there is any doubt as to the continuity between sections of floor trunking it is advisable to run an insulated protective conductor in the trunking. Protective conductors must connect from the trunking to earthing terminals of socket outlets and other accessories. Where socket outlets are required in positions where there is no floor or skirting trunking, such points can be wired in conduit connected to the side of the trunking.

Another type of metal floor trunking is the 'In-slab' installation method. This consists of enclosed rectangular steel ducts (usually 75mm × 35mm), together with junction and outlet boxes. A separate duct is provided for each wiring system, i.e. for low-voltage circuits, fire alarms, telephone lines, etc.

The separate ducts are spaced apart to give a stronger floor slab. The depth of the trunking and outlet boxes together with their supporting brackets equals that of the floor structure, so there is no need for a finishing screed, thus affording a considerable saving in construction costs. The outlet boxes can be fixed in any position, but a distance of 1.5 m between boxes will usually provide facilities for most office needs.

10.3 NON-METALLIC TRUNKING

A number of versatile plastic trunking systems have been developed in recent years and these are often suitable for installation work in domestic or commercial premises, particularly where rewiring of existing buildings is required. The trunking can be surface mounted and if care is taken in the installation, it can be arranged to blend unobtrusively into the decor. Skirting-mounted trunking is probably the most appropriate for use in domestic dwellings, but shallow multi-compartment trunking can also be run at higher levels in, for example, school classrooms or kitchens. Industrial non-metallic trunking is also available in a range of sizes up to 150mm × 150mm. The manufacturers of plastic trunking generally supply a full range of fittings and accessories for their systems, and in some cases these are compatible between one make and another. Generally, however, once one system is chosen, it will be necessary to stay with it to achieve a neat appearance and the ability to inter-change fittings.

The IEE Regulations which apply to metal trunking also generally apply to non-metallic types. Low-voltage insulated or sheathed cables may be installed in plastic trunking. In any area where there is a risk of mechanical damage occurring, the trunking must be suitably protected. Being non-conductive, it will be necessary to run protective conductors for circuits requiring them inside the trunking, and the size of these protective conductors must be calculated so as to satisfy the IEE Wiring Regulations.

The advantages of non-metallic trunking are that it is easier to install, is corrosion resistant and is maintenance free. In addition the flexibility is such that it is often possible to reposition outlets or make other alterations without any major disturbance. For those circumstances where it is required, plastic trunking can be obtained with metal screening between the different compartments used for low voltage, communication or other cables. There are limits to the ambient temperature in which the system can be installed.

FIGURE 10.4 Multi-compartment skirting trunking allows the segregation of different types of circuit. In this example 13 A ring main socket outlets and telecommunications circuits are provided (W.T. Parker Ltd).

Installation of Non-Metallic Trunking

Care and good workmanship are needed to ensure a successful installation, and the use of good quality materials is necessary. The installation layout must be planned before commencing work. If the installation is in a new or altered building, all internal structural and wall finishes should have been completed.

As with plastic conduit, it is necessary to allow for expansion of the trunking. This is done by leaving gaps between trunking sections as they are installed. A gap of 4–6mm per 3 m length is recommended if high ambient temperature variations are likely to occur. The gaps are generally covered by pieces designed for the purpose. The detail will vary according to the particular system being used and the manufacturer will be able to advise on the recommended method.

The trunking should be cut using a fine tooth saw. Clean off any burrs and swarf after making the cut. Appearance will be spoiled if the cut angles do not match exactly so it is advisable to use a mitre box to make the cuts.

The main component of the trunking is generally fitted to the surface of the wall using dome-headed screws. It is essential to use washers under the screw heads, and to cater for expansion of the plastic components, oversize holes should be drilled in the trunking. Trunking should be fixed at intervals of not more than 500mm, and there should also be fixings within about 100mm of the end or of any joint. If it is intended to fit any load-bearing components such as light fittings, extra fixings should be provided. It is best to first drill the clearance holes in the trunking, and then use the prepared length as a template to mark the wall for drilling. It is possible to use shot-fired masonry pins to

FIGURE 10.5   A typical office installation where data, telecommunication and power circuits are required (W.T. Parker Ltd).

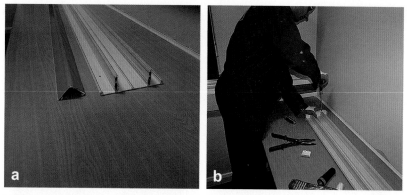

(a) Three-compartment dado trunking ready for fitting. The compartments and one of the lid sections can be clearly seen. (b) The end caps are screwed into position.

(c) Having checked the length required, cut the trunking using a fine-tooth hacksaw, the complete section is screwed in position on the wall. In this case the battery powered electric screwdriver is fitted with a light to aid the work. (d) The flat twin and cpc cable is prepared for use by running it out to avoid twists and kinks.

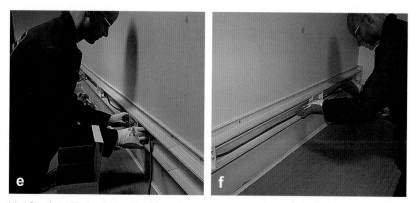

(e) After the cable has been placed into the centre compartment, the socket boxes are clipped in position. (f) The cable feeding the socket outlets is installed behind the outlet boxes.

FIGURE 10.6 Installing multi-compartment dado trunking.

Continued

(g) The lids are cut to length and fitted and butted up to the socket boxes. (h) In a similar way, the top and bottom lids are cut and fitted. These compartments will be used for data and communications circuits. The power cabling is complete and ready for the sockets to be wired.

(i) After work is complete, the site is left tidy, removing all rubbish and vacuum cleaning the floor (all M.W. Cripwell Ltd).

FIGURE 10.6   cont'd.   Installing multi-compartment dado trunking.

secure the trunking if desired. In this case it is essential to use cushioning washers under the heads of the pins.

In general, the various components of trunking systems clip together, but it may be necessary with some systems to employ glued joints. Special solvent adhesives are available for this purpose and should be applied in the same way as described in the section on installation of plastic conduit.

Once the trunking has been fixed, the cables can be run. Some makers supply special cable retaining clips which make it easier to retain cables prior to fitting the lids. Alternatively, it is a good idea to use short offcut sections of trunking lid for this purpose. Cable capacities are calculated in the same way as for conduit using a 'unit system'. The manufacturer of the trunking should be consulted for factors for other shapes.

When fitting the trunking compartment lids, increased stability and improved appearance will be achieved if the lid joints are arranged *not* to coincide with the joints in the main carrier.

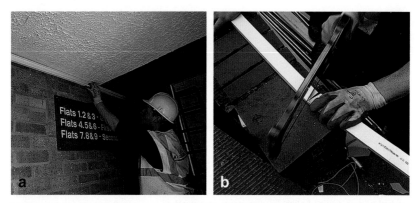

(a) The mini-trunking is offered up on site and marked to indicate the location of the bend. (b) Using a fine-tooth hacksaw, the conduit is cut at the back and on one side to suit the angle of bend required.

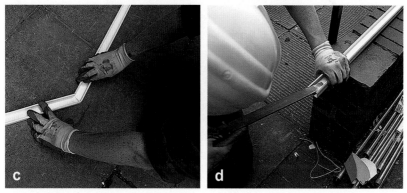

(c) The bend is tested on the ground prior to being offered up on the wall. (d) After marking and cutting to length, rough edges and burrs are removed using a file.

(e) Fixing holes are required and these are next drilled at suitable positions. (f) After drilling the conduit, the wall is correspondingly marked out and drilled for wall plugs.

FIGURE 10.7   Installing plastic mini-trunking.

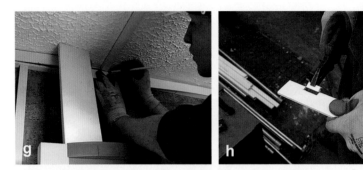

(g) The trunking lid is marked out for cutting. (h) After screwing the trunking in position, the lid is cut to suit. In this case a notch needs to be removed to clear an existing trunking run. The lid is notched using a saw and pliers used to remove the notch.

(i) The lid is fitted, the completed trunking gives a neat and workmanlike appearance (all M.W. Cripwell Ltd).

FIGURE 10.7   cont'd.   Installing plastic mini-trunking.

Skirting and Dado Trunking

Skirting and dado trunking is used extensively in commercial buildings, laboratories, hospitals and similar installations. It usually consists of a shallow PVC trunking, approximately 50mm deep with two or more compartments. One compartment is used for socket or lighting wiring, one for communications or telephone wiring, and very often a third compartment is reserved for data cabling to computers, as these cables must be separated from all other wiring systems.

Trunking can be shaped to form the skirting, and is frequently fitted around the outer walls of a building where sockets, telephones, etc., are likely to be required. It is often also fitted on internal walls. In order to cross the thresholds of doorways, and to interconnect isolated lengths of skirting trunking, conduits or floor trunking can be installed in the floor screed. Suitable bends and adaptors are made to connect between skirting and floor trunking.

Shallow flush-type socket-outlets can be obtained for fitting to the lid of skirting trunking and trunking manufacturers will punch suitable apertures for the reception of sockets.

FIGURE 10.8 (a) Underfloor three-compartment trunking installed in a commercial office installation. With the growth of data processing, flexible office wiring systems are a necessity, and a raised floor provides a viable method of achieving this. This outlet box is fitted with power sockets and data sockets will be fitted later. (b) The outlet box with the lid in position, providing a flat floor.

It is often an advantage to fit the sockets, data or telephone outlets on short lengths of lid, which need not be disturbed when the remainder of the lid is removed for extensions.

Another form of trunking in use is dado trunking incorporating busbars. These allow socket outlets and spur boxes to be simply plugged in, effecting an economy in installation times.

Where trunking passes through partitions, short lengths of lids should be fitted as this enables the remainder of the lid to be removed without difficulty.

Plastic Underfloor Trunking

As with many other types of wiring system available such as conduit or trunking, plastic materials are often used instead of their metal counterparts for the enclosures of underfloor systems.

Underfloor trunking systems made with this material can be divided into two main types, raised floor systems and underfloor ducted systems.

The raised floor installation has the advantage of extreme flexibility as the load-bearing floor is structurally supported such that there is an unobstructed space underneath. The wiring ducts can thus be run under the floor in any desired position. The outlet positions which are incorporated in floor panel sections are connected to the ducted wiring using flexible conduit and in this way outlet positions can be rearranged at will by exchanging the floor panel sections. This type of layout is especially useful in computer rooms where due to the rapid advance of technology it is necessary to replace obsolete equipment at intervals.

The other system supplied in plastic materials is the underfloor ducted system. With this, shallow ducts are installed prior to the final floor surface being laid. The ducting is subsequently buried in the concrete screed. A variety of outlet positions can be used. Concealed and raised socket outlets are available, and as previously mentioned, 'power poles' can also be fitted. Some manufacturers supply fittings whereby connection can readily be made to skirting trunking.

10.4 CABLE DUCTS

Cable ducting is defined in the IEE Regulations as 'an enclosure of metal or insulating material, other than conduit or cable trunking, intended for the protection of cables which are drawn in after erection of the ducting'.

Cable ducts usually consist of corrugated PVC, sometimes placed inside earthenware or concrete pipes buried in the slab or ground, with suitable access chambers to enable cables to be drawn in. IEE Regulation 522.8.3 requires that every bend formed shall be such that cables will not suffer damage. Cables installed in underground ducts should have a sheath or armour to resist any mechanical damage. Unsheathed cables must not therefore be installed in these ducts. Mineral insulated copper sheathed cables which are installed in ducts must have an overall covering of PVC sheath.

The space factor of ducts must not exceed 35%, whereas the space factor for trunking is 45%, and that for conduit is 40%. All of these space factors depend upon not more than two 90° bends (or the equivalent) being installed between draw-in points. IEE Regulation 528.1 makes it clear that Band I and Band II cables must not be installed in the same duct.

One method of forming concrete ducts is by means of a flexible rubber or plastic tubing of the required diameter. This is inflated and placed in position before the concrete slab is poured. After the concrete has set, the tube is deflated and withdrawn, and can be reused to form other ducts. Bends in ducts can be formed by this method provided the inner radius is not less than four times the diameter of the duct.

10.5 UNDERFLOOR TRUNKING SYSTEMS

Open plan office and other types of commercial buildings may well need power and data wiring to outlets at various points in the floor area. The most appropriate way of providing this is by one of the underfloor wiring systems available. Both steel and plastic construction trunkings can be obtained, and if required 'power poles' can be inserted at appropriate locations to bring the socket outlets to a convenient hand height. With the increasing use being made of computers and other electronic data transmission systems, the flexibility of the underfloor wiring can be used to good advantage.

Chapter 11

Busbar and Modular Wiring Systems

11.1 BUSBAR SYSTEM

The wide range of busbar system is available and can be used for single or three-phase distribution to many types of applications ranging from lighting to heavy-duty machines in factories. This consists of copper or aluminium busbars mounted on insulators and enclosed in standard lengths of steel trunking, which are arranged to be fitted together thus forming a continuous busbar along the entire length of the distribution route. It is sometimes more economical to use busbars in place of long runs of sub-main cables.

At intervals, typically every 1m, a tap-off point is provided. At these points tap-off units may be fitted; these can comprise unfused lighting units or power units protected with HRC fuses or circuit breakers.

The units are provided with contact fingers which are designed to fit onto the busbars.

Connections from these tap-off units to individual accessories, motors or other electrical equipment can be made by flexible connections, PVC sheathed cables, or conduit.

The advantages of this system are that the trunking and busbars can be erected before the installation of the machinery, and the latter can be connected up and set to work as soon as they are installed.

By bringing the heavy main feeders near to the actual loads, the circuit wiring is reduced to a minimum and voltage drop is lower than would otherwise be the case.

Subsequent additions and alterations to plant layout can be easily accomplished, and where busbar sections have to be removed they can be used again in other positions.

If a large number of small machines are to be fed it is usual to fit a distribution board near the trunking system and to protect this with a tap-off box fitted with HRC fuses of suitable capacity. Circuit wiring from the distribution board is usually carried out in heavy gauge screwed conduit.

The system is comparatively expensive in first cost and, therefore, is best employed where heavy loads and a large number of machines have to be provided for. However, whilst material costs may be higher, installation is quicker and overall there may be cost benefits from using the system. Once installed there is very little depreciation or need for maintenance, and it has a high recovery value should it be necessary to dismantle and install it elsewhere.

Several well-known manufacturers supply this trunking in various sizes, together with all necessary tees, bends, tap-off boxes and other accessories. Earth continuity is usually provided by an external copper earth link which ensures good continuity. Lighting trunking is also available whereby the trunking is fitted with integral busbars and the lighting fittings are simply clipped in place.

Where there are long runs of busbar trunking it is necessary to provide expansion joints to take up any variations in length due to changes in temperature. These expansion joints usually take the form of a short length of trunking enclosing flexible copper braided conductors instead of solid busbars. It is advisable to provide one of these in every 30m run of trunking. Busbars are available with additional auxiliary conductors and these may be used for emergency lighting or self-test systems. Busbars are available with seven conductors as standard and some have smaller auxiliary conductors. No other conductors of any kind may be installed inside trunking containing bare copper busbars.

All lids and covers must always be kept in position as a protection against vermin and also to avoid accidental contact with live busbars. The trunking should be marked prominently on the outside at intervals with details of the voltage of the busbars and the word DANGER. The conductors shall be installed so that they are not accessible to unauthorised persons.

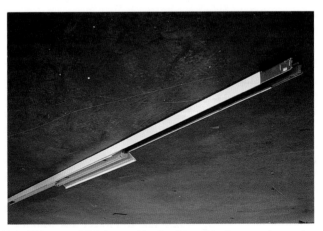

FIGURE 11.1 This type of lighting trunking has integral busbars carrying conductors for the supply to the luminaires. The 54-W twin fittings are available with either wide angle or semi-specular narrow beam characteristics, enabling effective lighting from different heights. The trunking can be suspended from the building frame and, with the narrow beam reflectors, is suitable for high-bay illumination. The clip-in fittings can be arranged to pick up from different phases and the whole assembly can be carried out without the use of tools, a particular advantage when working at heights (W.T. Parker Ltd).

FIGURE 11.2   The busbar lighting units installed in a high-bay storage building. The lighting units have been installed and the installation is ready for the lids to be clipped on (W.T. Parker Ltd).

Insulators shall be spaced to prevent conductors coming in contact with each other, with earthed metal or other objects. The conductors shall be free to expand or contract during changes of temperature without detriment to themselves or other parts of the installation. In damp situations the supports and fixings shall be of non-rusting material. If conductors are to be installed where exposed to flammable or explosive dust, vapour or gas, or where explosive materials are handled or stored, additional screens, caps and accessories should be fitted. These are able to improve the IP rating of the equipment.

Where the trunking passes through walls or floors no space shall be left round the conductors where fire might spread. Fire barriers should be provided at these points inside the trunking.

All runs of overhead busbar trunking must be capable of isolation in case of emergency or maintenance by means of an isolating switch fixed in a readily accessible position (Electricity at Work Regulations 1989 – Regulation 12).

Busbar System for Rising Mains

A similar busbar system is frequently used for vertical rising mains for multi-storey buildings. This usually consists of copper or aluminium busbars of capacities of 100–2500A with two, three or four conductors. These are usually metalclad and are made in 4m sections, although all-insulated rising busbar systems are also obtainable.

Tap-off boxes with fuselinks or fuseswitches can be provided for distribution to each floor where distribution boards can be fitted near the tap-off units. For these vertical runs it is very important that fire-resisting barriers be fitted inside the trunking at the level of each floor. These fire barriers can be purchased with the trunking, and the manufacturers will fit these in the required positions if provided with the necessary details. Where the trunking passes through floors,

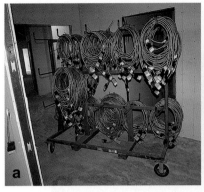
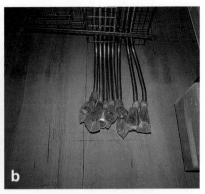

(a) The pre-formed wiring units with plugs attached are conveniently stored and carried on this wheeled trolley preparatory to installation. All units are numbered to ensure correct installation. (b) Modular wiring installed ready for the distribution board to be secured and plugged in.

(c) Distribut ion board in position with modular wiring looms fitted into the distribution cupboard, carried on a cable basket route. (d) Installing modular cable units on cable basket behind a suspended ceiling.

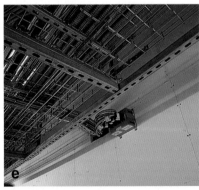

(e) The junction box and cable basket enable a neat installation to be made and one where fault finding and alterations are easily accommodated at a later date. (f) Modular units installed in a wall partition. Sufficient spare cable is arranged to enable the fitting of switch and socket units and to carry out changes should these be needed at a later date (all M.W. Cripwell Ltd).

FIGURE 11.3　Modular wiring.

whether in a specially formed riser cupboard, or run on the surface of a wall, it is necessary to ensure that the floor is 'made good' by non-combustible material round the outside of the trunking to prevent the spread of fire.

The supply for these runs of busbars is usually effected by a feeder box with provision for whatever type of cable is used. The manufacturers should be consulted as to the correct size of busbars to use, and IEE Regulations recommend that the maximum operating temperature should not exceed 90 °C. Where rubber or PVC cables are connected to busbars operating at comparatively high temperatures the insulation and sheath must be removed for a distance of 150mm from the connection and replaced by suitable heat resisting insulation.

11.2 MODULAR WIRING SYSTEM

Modular wiring is encountered in large sites where multiple installations of identical or similar design are required, or where speed of on-site installation is paramount in progressing the installation work. Modular wiring comprises factory assembled units with cables bunched together inside flexible conduit, and with end plugs pre-wired ready for assembly to matching sockets fitted to the equipment on-site. The units are made up in pre-arranged lengths with the cores of appropriate cross-section connected to the correct pins in the plugs. 'Home runs' are used where the cross-sectional area of the cables is arranged to take the load of several circuits and these feed junction units from which modular final circuits are run.

Whilst the initial cost of the equipment may well be greater than with traditional wiring, one main advantage of the system is that all the connections are made up in controlled conditions in a factory where the working environment is more stable than on a building site. A second advantage is that installation on-site is quick and easy. Once the cable containment is complete (typically using cable tray, basket or ladder), it is a matter only of selecting the correct pre-assembled modular cable, running it on the correct route and plugging in the end fittings to the on-site equipment plugs. It is, of course, important to select the correct 'loom' for the individual application, these being identified by numbering during manufacture. Routing in rooms is usually inside the wall void and large diameter holes (100mm) allow the end plugs to be connected to individual switches, outlets or accessories. Any surplus length of the modular unit is fed back into the wall void once the switch or socket is fitted. This can then easily be recovered if any extensions or alterations are later required.

Chapter 12

Power Cable Systems

A range of cable types is available, and these include PVC insulated, XLPE insulated and LSF type, usually with wire armouring. Each has characteristics which can be appropriate to a range of installation situations, and some detail is given in the sections which follow. In the past paper-insulated lead-covered (PILC) cables were used extensively for power cabling and may occasionally be encountered in old installations.

12.1 ARMOURED, INSULATED AND SHEATHED CABLES

Armoured XLPE and PVC insulated cables are used extensively for main cables and distribution circuits, and also for circuit wiring in industrial installations.

These cables consist of multi-core PVC insulated cables, with PVC sheath and steel wire armour (SWA), and PVC or XLPE sheathed overall. The main disadvantages of PVC insulated cables are that thermo-plastic insulation will sustain serious damage if subject to temperatures over 70°C for a prolonged period, and proper protection against sustained overloads is required. The insulation will harden, and become brittle in temperatures below 1 °C, and the cables should not be installed or handled when temperatures are approaching freezing, otherwise the insulation may be inclined to split. Low temperatures will do no permanent harm to the insulation, providing the cables are not interfered with during extreme cold. PVC/SWA/PVC multi-core sheathed cables are manufactured in all sizes up to 400mm². The latter are heavy to handle and threading through the cable route can be difficult. This can be avoided in some cases by installing two parallel cables. It may well be possible to replace one 400mm² with two 140mm² cables, with a cost saving and easier installation. Design details for parallel cables are given in Chapter 4.

Details of current ratings are given in the IEE Cable rating tables. These cables can be laid directly in the ground, in ducts, or fixed to the surface on a cable tray, or fixed to the structure by cleats. When a number of multi-core cables take the same route, it is an advantage for them to be supported on cable trays or ladders, which are manufactured in various sizes from 50mm to 1200mm wide (see Section 'Cable tray, cable basket and cable ladder' below). When several cables are grouped together on a wall, or tray, or in ducts, the

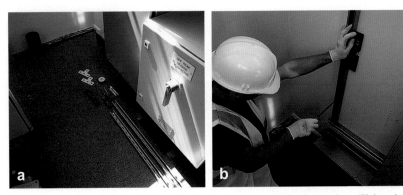

(a) A new incoming main switch is to be installed using metal framing such as 'Unistrut' to support the switchgear, the components for which are laid out ready. (b) After marking out the required positions, the metal is cut to length. Secure wall fixings must be found and trial holes are sometimes needed in a studded wall to find the structural supports.

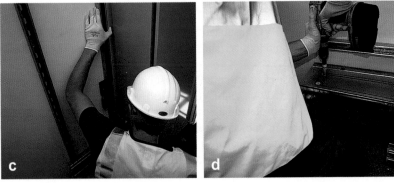

(c) Here the second strut is being screwed into position. (d) The trunking requires an access slot and after marking out, the corners are bored out using a hole-saw.

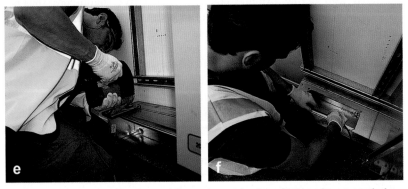

(e) The slot is cut using an electric saw. After cutting out the slot, a file is used to remove the burrs and prepare a smooth edge. (f) A grommet strip if fitted to ensure that damage to cables is prevented when they are installed.

FIGURE 12.1   Installation of main incoming switch.

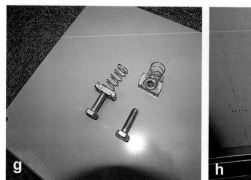
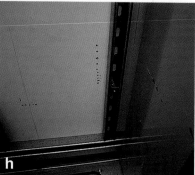

(g) The vertical members allow universal fixing and spring clips, known colloquially as 'zebs' are fitted to accept the fixing screws. The name derives from a childrens' TV programme featuring a character called zebedee. (h) The metal framing with the 'zebs' in position ready to receive the fixing bolts.

(i) The main switch bolted in position. The slot in the top of switch housing is ready to accept distribution panel, the next stage of the installation.

FIGURE 12.1   cont'd.   Installation of main incoming switch.

current rating will have to be reduced according to the correction factors as described in Chapter 4.

The smaller multi-core cables have many advantages when used in industrial installations for circuit and control wiring owing to their ease of installation, flexibility, and high recovery value when alterations become necessary.

End terminations are made by stripping back the PVC sheathing, and steel wire armouring, and fitting a compression gland which can be screwed to switchgear, etc., and provide earth continuity between the armour of the cable and the switchgear. When connecting to motors on slide rails, a loop should be left in the cable near the motor to permit the necessary movement.

Single-core cables armoured with steel wire or tape shall not be used for a.c. (IEE Regulation 521.5.2), but single-core cables with aluminium sheaths may

FIGURE 12.2   Cable runs fitted in the roof space above a suspended ceiling. Multi-compartment cable basket is provided, as well as cable tray for different circuit categories. Note the copex flexible conduits which carry feeds to specific locations (W.T. Parker Ltd).

be used provided the current ratings in IEE Table 4H1 are complied with, and that suitable mechanical protection is provided where necessary. Single-core cables are not usually armoured.

Metal armouring of PVC/SWA cables which come into fortuitous contact with other fixed metalwork shall either be segregated therefrom or effectively bonded thereto. PVC/SWA armoured cables shall have additional protection where exposed to mechanical damage; for example, cables run at low levels in a factory and might be damaged by a fork lift truck. To cater for damp situations, and where exposed to weather, suitably rated cable glands can be obtained and these will improve upon the IP rating of standard glands. The metal armouring of cables should be of corrosion-resisting material or finish, and must not be placed in contact with other metals with which they are liable to set up electrolytic action. This also applies to saddles, cleats and fixing clips (IEE Regulation 522.5).

If it becomes necessary to carry out any work on multi-core cables that have already been in service, precautions must be taken to ensure that no current is present in the cable due to its capacitance. Long runs of cable act as capacitors when in service, and when disconnected from the source of supply a high potential may have been built up in the cable. Before touching any of the conductors, therefore, the current should be discharged by connecting a lamp, resistor or voltmeter between earth and each conductor in turn.

In the case of underground multi-core cables that have been in service and have to be cut, it is usual to spike the cable with a metal spike at the position where it is proposed to cut it. The spike should penetrate the earthed sheath and all the conductors; this will ensure that the cable is discharged, and will also obviate any risk of accidentally cutting through a cable that may be live.

XLPE Cables

XLPE cables are made to BS 5467 and are virtually standard for use in new installations. XLPE has better insulation qualities than PVC and thus it is possible to obtain cables of a smaller diameter for the same voltage rating. In addition, provided suitable terminations are used, XLPE cable may be used at a maximum working temperature of 90°C and they can be installed at temperatures as low as −30 °C should the need arise. These cables are available in sizes up to 400mm^2 or 1000mm^2 single core.

Jointing must be given careful consideration, and crimping is recommended rather than soldering if advantage is to be taken of the maximum short-circuit capability of the cable. Also, the jointing compound used must be selected to

FIGURE 12.3   Sub-main distribution cables run on cable ladder and cleated. The PVC/SWA/ PVC cables shown are four-core 240mm^2 and feed four switchboards in a hospital. Each switchboard has separate feeds for essential and non-essential supplies (hence eight cables), and in the event of mains failure, the essential circuits are fed by diesel driven emergency alternators (William Steward & Co. Ltd).

suit XLPE, and some materials such as PVC tape are incompatible and must not be used. Compounds which suit LSF as well as XLPE are available.

LSF and LSOH Sheathed Cables

In situations where there is a need to protect people who may be at risk due to the outbreak of fire, low smoke and fume (LSF) cables made to BS 6724 may be used. In addition, low smoke zero halogen (LSOH) cables are available. In certain circumstances, emissions of toxic gases are reduced, and the cables are slow to ignite. This reduces the risk to occupants, and increases the ability to escape.

Locations where this may be relevant include underground passageways or tunnels, cinemas, hospitals, office blocks and other similar places where large numbers of people may be present. Cables are available in sizes up to 630mm^2 single core, and 400mm^2 two, three or four core. Typical cables have copper conductors, XLPE insulation with LSF bedding, single wire armouring and an LSF or LSOH sheath. Control cables with LSF insulation can also be obtained in sizes up to 4mm^2 and with up to 37 cores.

Paper-Insulated Lead-Covered Cables (PILC)

Paper-insulated lead-covered cables are seldom used today but were formerly used extensively for power cabling and may occasionally be encountered in old installations.

Paper is a good insulator but loses its good properties if it becomes damp and thus the terminations and joints should be protected from the ingress of moisture by being suitably sealed. Due to the special skills required for jointing these cables they have been replaced by PVC or XLPE insulated and armoured cables. If they need to be manipulated, the maximum internal radii of bends in PILC cables shall not be less than 12 times the overall diameter of the cable.

12.2 CABLE TRAY, CABLE BASKET AND CABLE LADDER

Cable tray, cable basket and cable ladder are ideal methods of supporting cables in a variety of situations. With care, a very neat appearance can be obtained, and with both vertical and horizontal runs, cables can be run in line, free from deviations round beams or other obstructions. In buildings with suspended ceilings, cable tray or basket offers an ideal method of running wiring in the ceiling void.

Cable tray comprises the basic lengths of galvanised steel tray, usually in 3m sections, and a range of fixings to enable the sections to be joined and run round vertical or horizontal bends. Various support accessories are also available, and the accompanying illustrations show the steps needed in the assembly

(a) Supports appropriate for the installation are first fitted. Here ceiling supports to carry cable tray and trunking above a suspended ceiling can be seen. (b) The length of tray required is measured and marked off, ready for cutting.

(c) The cable tray is cut using a hacksaw. (d) Using a file, burrs are removed to prevent damage to cables.

FIGURE 12.4 Installation of Cable Tray.

Continued

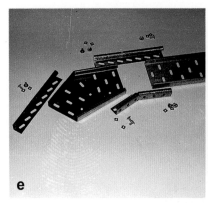

(e) For a change of direction, a pre-formed bend can be used. The assembly is bolted in position using readily obtainable joining strips, one of which has been cut on one flange only and bent to suit the angle required. Here are the components of the bend, ready for assembly. (f) The individual components are assembled with nuts and bolts, here seen being tightened. Cable tray is often run in inaccessible positions and so as much work as possible should be carried out on the bench.

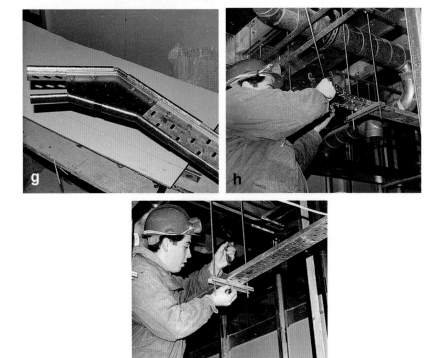

(g) The assembled bend, ready to be erected into the run of cable tray. (h) Here the bend has been placed in position, and a measurement is being taken for the next straight length prior to cutting. (i) Finally, the cable tray run is secured to the support bars, again using nuts and bolts, assembled with the head uppermost (all William Steward & Co. Ltd).

FIGURE 12.4   cont'd.   Installation of Cable Tray.

FIGURE 12.5 A neat installation of cable tray in a sub-station carrying feeds to individual distribution boards (W.T. Parker Ltd).

FIGURE 12.6 Cables connecting this standby alternator are mounted on cable tray.

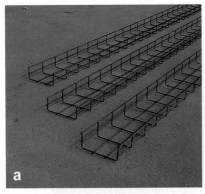
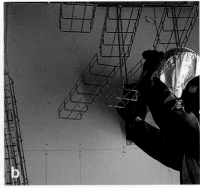

(a) Straight lengths of cable basket are available in a range of sizes, with and without a barrier to segregate different cable groups. These 100 x 50mm sections are ready for fitting. (b) In this installation there are three parallel runs being erected and, as part of co-ordination of site services, will need to be offset to avoid the adjacent ventilation ducting. Here, the initial cuts have been made and the bends for two of the three runs are positioned.

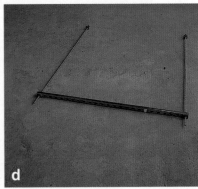

(c) The next stage is to secure the inner edge of each bend using bolted clips. (d) To support the basket runs from the ceiling, suspension brackets are prepared.

(e) The suspension brackets are fitted in position and a spirit level is used to check that the spars are truly horizontal. (f) To complete the third run of cable basket, the offset can be assembled on the floor. All cuts are made using the bolt cutters as shown, initially to the side members of the run. By cutting the elements at a suitable angle, sharp edges to the cuts can be avoided.

FIGURE 12.7   Installation of Cable Basket.

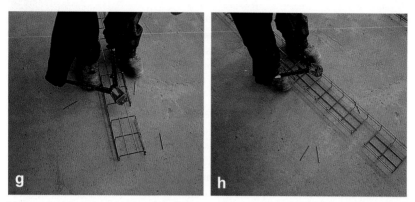

(g) Two cuts are needed to remove each of the base elements of the basket. (h) The second bend of the offset is dealt with by cutting in the same way.

(i) Once the side and bottom elements have been cut, the bend can be made and secured with the bolted clips. (j) The completed offset, ready for erection overhead into the third basket run.

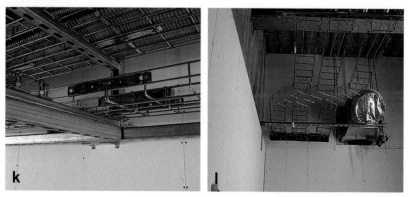

(k) In-line joints may be bolted or clipped. In this view, the clipped joining strips are seen. (l) The completed set of three parallel runs of cable basket, with suitable openings cut in the partition and offset to allow for the later erection of the ventilation ducting.

FIGURE 12.7 cont'd. Installation of Cable Basket.

of a typical run. Cable tray is cut using a hacksaw and after cutting the burrs are removed with the use of a file. The sections are joined using joining strips which are bolted in place with galvanised bolts and nuts. Some makes of tray may be joined by the use of spring clips. Light duty tray may be bent by hand after cutting the side flange, and special bending machines deal with bending of heavy-duty cable tray.

Light and heavy section cable tray is available, and for some of the heavier duty types, there can be an increased distance between supports. The use of such trays enables considerable savings to be made in fixing costs.

Cable basket is similarly obtained in fixed lengths and is available in a range of widths. Neat and efficient containment with both vertical and horizontal runs can be made. It is also possible to obtain sections with a metallic division to achieve segregation of circuits where this is needed. Cable basket is generally cut using 'bolt cutters' and with care and practice it is possible to make up neat vertical and horizontal bends. The lengths of basket may be bolted or clipped together, depending on the design. If a high integrity of containment is required, the joints may be welded. As with any other form of cable containment, the assembly must be completed before cable runs are put into position (Figure 12.7).

Sheathed and/or armoured cables which are run on cable trays need not be fixed to the tray provided the cables are in inaccessible positions and are not likely to be disturbed, and that the cables are neatly arranged in such a manner that the route of each cable can be easily traced. With large cables, it is necessary for the cables to be secured, as there can be a significant electro-mechanical stress should a high fault current arise.

In many industrial buildings the roof purlins are specially shaped to accommodate multi-core and other types of cable, thus eliminating the need for any additional method of support or fixing. The 'Multibeam' system comprises purlins specially designed to accommodate cables, and various accessories are available to provide for outlets for lighting fittings and power points. Where unsheathed PVC cables are installed in these purlins, insulated covers are provided to give the necessary protection against mechanical damage.

Chapter 13

Insulated and Sheathed Cable Systems

The insulated and sheathed system is used extensively for lighting and socket installations in small dwellings, and is probably the most economical method of wiring for this type of work. It will be appreciated that the amount of mechanical protection provided to the cables is limited and care should be taken to avoid situations where this would introduce risk. It is customary to use two- and three-core cables with an integral protective conductor and to provide insulated joint boxes or four-terminal ceiling roses for making the necessary connections. An alternative method of wiring with PVC-sheathed cables for lighting is to use two-core and c.p.c. cables with three-plate ceiling roses instead of joint boxes.

IEE Regulation 526.5 requires that terminations or joints in these cables must be enclosed in non-combustible material, such as a box complying with BS 476 part 12, or an accessory or luminaire. (An 'accessory' is defined as 'a device, other than current-using equipment, associated with such equipment or with the wiring of an installation'.)

At the positions of joint boxes, switches, sockets and luminaires the sheathing must terminate inside the box or enclosure, or could be partly enclosed by the building structure if constructed of incombustible material.

13.1 SURFACE WIRING

When cables are run on the surface a box is not necessary at outlet positions, provided the outer sheathing is brought into the accessory or luminaire, or into a block or recess lined with incombustible materials, or into a plastic patress.

For vertical-run cables which are installed in inaccessible positions and unlikely to be disturbed, support shall be provided at the top of the cable, and then at intervals of not less than 0.5m. For horizontal runs the cables may rest without fixings in positions which are inaccessible and are not likely to be disturbed, provided that the surface is dry, reasonably smooth and free from sharp edges. The minimum radii of bends in PVC cable are specified. For those of 10mm diameter or under, bends must be at least three times the diameter; for up to 25mm, bends should be at least four times the diameter and for over 25mm, six times.

FIGURE 13.1 A kitchen ring main being installed in a dwelling house using concealed wiring. Steel capping has been secured in the permitted positions (Fig. 13.7) and the outlet boxes have been secured and cable installed prior to plastering taking place (NBK Electrical).

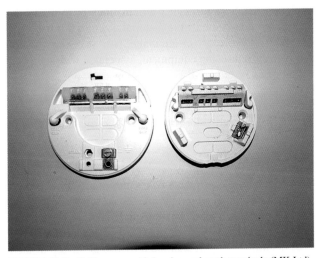

FIGURE 13.2 Ceiling rose with looping and earth terminals (MK Ltd).

PVC and similar sheathed cables if exposed to direct sunlight shall be of a type resistant to damage by ultraviolet light (IEE Regulation 522.11). PVC cables shall not be exposed to contact with oil, creosote and similar hydrocarbons, or must be of a type to withstand such exposure (IEE Regulation 522.5).

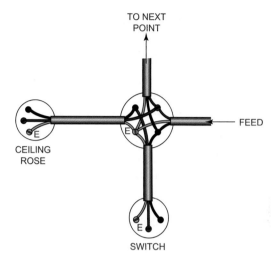

FIGURE 13.3 PVC-sheathed wiring system. Joint box connections to a light controlled by a switch, with cable colours indicated.

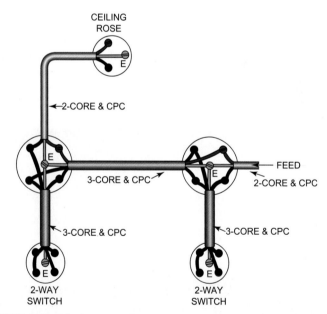

FIGURE 13.4 Joint box connections to two two-way switches controlling one light.

13.2 CONCEALED WIRING

PVC wiring, concealed in floors or partitions, is an effective method of providing a satisfactory installation where appearance is of prime importance as in domestic, display or some office situations. Such wiring arrangements are covered by IEE Regulation 522.6.6. There is no reason why PVC-sheathed cables shall not be buried directly in cement or plaster, provided the location is such that IEE Regulation 522.6.6 is complied with and RCD protection in accordance with IEE Regulation 522.6.7 is provided. The cable locations permitted by IEE Regulation 522.6.6. are illustrated in Fig. 13.7. A disadvantage is that cables once buried in cement or plaster cannot be withdrawn should any defect occur, and the circuits would then have to be rewired. It is better to provide a plastic conduit to the switch or outlet positions so that the PVC cables can be drawn into the conduit, and withdrawn should the need arise. Such an arrangement must also comply with the location constraints given in Fig. 13.7. The RCD protection may be omitted if the cables are enclosed in steel conduit or under capping capable of resisting penetration by nails or screws.

If it is impractical to run concealed wiring in the location zones specified, then appropriate protection must be provided. This may take the form of a cable incorporating an earthed metal sheath, or by enclosing the cables in earthed metallic conduit, trunking or ducting. The addition of RDC protection is a requirement irrespective of cable location if the wall or partition includes any metallic components.

Whichever construction is employed, it is necessary to provide a box at all light, switch and socket outlet positions. Metallic boxes must be provided with earthing terminals to which the protective conductor in the cable must be connected. If the protective conductor is a bare wire in a multi-core cable, a green/yellow sheath must be applied where the cable enters the box (IEE Regulation 514.3.2).

FIGURE 13.5 During building construction, it is possible to gain access to joists from either above or below. Here the upper floor has been laid but access for wiring is still possible from below. Joists have been bored and cable runs made prior to the ceiling being fitted (NBK Electrical).

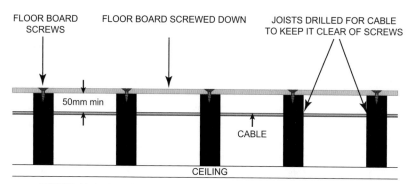

FLOOR BOARD   FLOOR BOARD SCREWED DOWN   JOISTS DRILLED FOR CABLE
SCREWS   TO KEEP IT CLEAR OF SCREWS

50mm min

CABLE

CEILING

FIGURE 13.6 Running PVC-sheathed cable under wooden floors across joists.

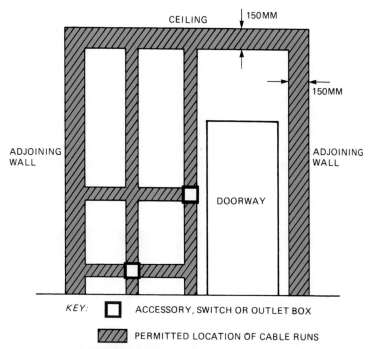

CEILING   150MM

150MM

ADJOINING   ADJOINING
WALL   WALL

DOORWAY

KEY: □   ACCESSORY, SWITCH OR OUTLET BOX

           ▨   PERMITTED LOCATION OF CABLE RUNS

FIGURE 13.7 Typical permissible locations for concealed cable runs (IEE Regulation 522-06-06).
Where it is impractical to use these locations, special precautions are necessary, see text.

Keep Cables Away From Pipework

Insulated cables must not be allowed to come into direct contact with gas pipes
or non-earthed metalwork, and very special care must be exercised to ensure
that they are kept away from hot water pipes.

Precautions Where Cables Pass Through Walls, Ceilings, etc.

Where the cables pass through walls, floors, ceilings and partitions, the holes shall be made good with incombustible material to prevent the spread of fire. It is advisable to provide a short length of pipe or sleeving suitably bushed at these positions, and the space left inside the sleeve should be plugged with incombustible material. Where the cables pass through holes in structural steelwork, the holes must be bushed so as to prevent abrasion of the cable.

Where run under wood floors, the cables should be fixed to the side of the joists, and if across joists, should be threaded through holes drilled through the joists in such a position as to avoid floorboard nails and screws.

Where cables are sunk into floor joists the floorboards should be fixed with removable screws. In any case, screwed 'traps' should be left over all joint boxes and other positions where access may be necessary.

Wiring to Socket Outlets

When PVC cable is used for wiring to socket outlets or other outlets demanding an earth connection it is usual to provide two-core and c.p.c. cables. These consist of two insulated conductors and one uninsulated conductor, the whole being enclosed in the PVC sheathing. It is necessary to check that the protective conductor complies with IEE Section 543.

When wiring to 13Amp standard domestic sockets, the cables will have to be taken into the box which is designed for these sockets and which includes an earth terminal.

(a) The sheathing is stripped back to allow connection of the cores to the socket outlet. This must only be removed as far as necessary to enable the insulated conductors to be manipulated. (b) Side-cutters are used to cut off unwanted part of the sheath.

FIGURE 13.8   Wiring to socket outlets.

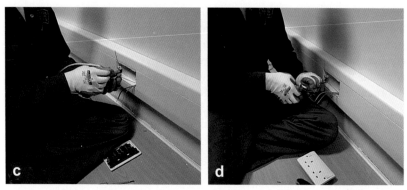

(c) A green-yellow sleeve is slid onto the bare protective conductor and cut to length. (d) Insulation is stripped from the ends of the live and neutral conductors, just sufficient to make effective contact when placed into the socket terminal tunnel.

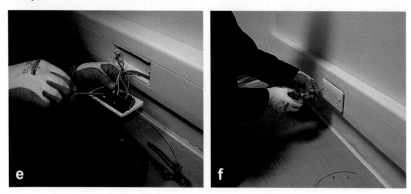

(e) The conductors are connected to the correct terminal tunnels of the socket, ensuring the screws are properly tightened to make for good conductivity and to ensure they do not subsequently become detached. Socket outlets with tunnel type terminals are preferred as these terminals enable maximum and uniform pressure to be applied on up to two main circuit cables and one spur cable. (f) After all terminal connections have been made, the slack cables should be carefully disposed to avoid cramping. Finally the socket outlet may be pushed gently into the box and secured by the two fixing screws (all M.W. Cripwell Ltd).

FIGURE 13.8 cont'd. Wiring to socket outlets.

Practical Hints

Care must be taken when stripping the sheathing of PVC cables so as to avoid nicking the inside insulation.

The sheathing must only be removed as far as necessary to enable the insulated conductors to be manipulated and connected. The sheathing must be taken well into junction boxes, switch boxes, etc., as the insulation must be protected over its entire length.

Multi-core cables have cores of distinctive colours; the brown should be connected to phase terminals, the blue to neutral or common return, and the protective conductor to the earth terminal. Clips are much neater than saddles, but when more than two cables are run together it is generally best to use suitable insulated saddles. If a number of cables have to be run together on concrete or otherwise in locations where the fixings are difficult to obtain, it is advisable to fix a wood batten and then to clip or saddle the cables to the batten. Information on the spacing of fixings for horizontal and vertical cable runs is given in the IEE On-site Guide (Fig. 13.9).

Cable runs should be planned so as to avoid cables having to cross one another, and additional saddles should be provided where they change direction. PVC-sheathed cables should not be used for any systems where the normal voltage exceeds 1000V.

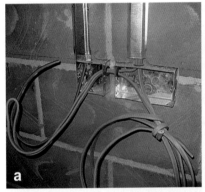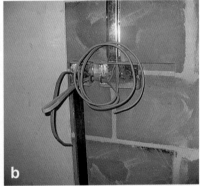

(a) After the position of cable drops is decided, the blockwork is chased out to the required depth and the boxes fitted. Cabling is run and steel capping secured to protect from mechanical damage. (b) Plasterboard is fitted in position prior to the final surface being applied. Note that the cable runs in this installation are vertical and comply with the positions shown in Fig. 13.7.

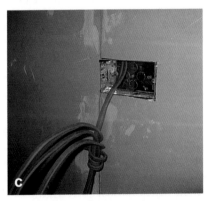

(c) After the wall surface is complete, the cables can be stripped and socket outlets, spur boxes and other accessories fitted to complete the installation (all NBK Electrical).

FIGURE 13.9 Installing concealed wiring.

Chapter 14

Installation of Mineral Insulated Cables

Mineral insulated (MI) cables have been in use for a sufficient number of years to have stood the test of time. These cables have an insulation of highly compressed magnesium oxide powder (MgO) between cores and sheath. During manufacture the sheath is drawn down to the required diameter; consequently the larger sizes of cable yield shorter lengths than the smaller sizes. Generally MI cable needs no additional protection as copper is corrosion resistant. However, in certain hostile environments, or if a covering is required for aesthetic or identification purposes, MI cable is available with PVC or LSF covering.

The advantages of MI cables are that they are self-contained and require no further protection, except against the possibility of exceptional mechanical damage; they will withstand very high temperatures, and even fire; they are impervious to water, oil and cutting fluids, and are immune from condensation. Being inorganic they are non-ageing, and if properly installed should last almost indefinitely.

The overall diameter of the cable is small in relation to its current-carrying capacity, the smaller cables are easily bent and the sheath serves as an excellent protective conductor. Current-carrying capacities of MI cables and voltage drops are given in IEE Tables 4G1A and 4G2A.

14.1 FIXING

The cable can be fixed to walls and ceilings in the same manner as PVC insulated cables. A minimum bending radius, of six times the bare cable outside diameter is normally applicable. This permits further straightening and re-bending when required. If more severe bends are unavoidable, they should be limited to a minimum bending radius of three times the bare cable diameter, and any further straightening and bending must be done with care to avoid damaging the cable.

All normal bending may be carried out without the use of tools, however, two sizes of bending levers are available for use with the larger diameter cables or when multiple bends are required. These levers are specially designed to prevent cable damage during bending.

FIGURE 14.1   Control wiring in $3 \times 1.5mm^2$ PVC-sheathed MI cable (Wrexham Mineral Cables).

FIGURE 14.2   Straightening up multiple runs of MI cable using a block of wood and hammer (Wrexham Mineral Cables).

When carrying out the installation of this wiring system, the sheathing of the cable must be prevented from coming into contact with wires, cables or sheathing or any extra-low voltage system (not exceeding 50V a.c. or 120V d.c.), unless the extra-low voltage wiring system is carried out to the same requirements as for a low-voltage system (1000V a.c.). This means that it must not be allowed to come into contact with lightly insulated communication or data cables.

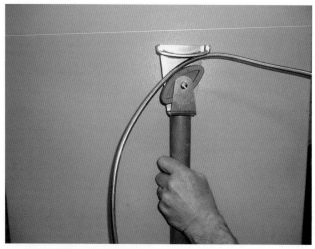

FIGURE 14.3   Bending and setting of MI cable. These operations can be more easily done by means of the simple tool illustrated (Wrexham Mineral Cables).

Protection Against Mechanical Damage

Mineral insulated cable will withstand crushing or hammering without damage to the conductors or insulation. However, if the outer sheathing should become punctured, the insulation will begin to 'breathe' and a low insulation resistance will result. Therefore, it is advisable to protect the cable if there is a possibility of its being mechanically damaged.

Where cables are exposed to possible mechanical damage it is advisable to thread the cables through steel conduits, especially near floor levels, or to fit steel sheathing over the cables in vulnerable positions. Where cables pass through floors, ceilings and walls the holes around the cables must be made good with cement or other non-combustible material to prevent the spread of fire, and where threaded through holes in structural steelwork the holes must be bushed to prevent abrasion of the sheathing.

14.2 BONDING

Because of the compression-ring type connection between the gland and the cable, and the brass thread of the gland, no additional bonding between the sheath of the cable and connecting boxes is necessary although it is common practice for the terminating pot to have an earth tail fitted. The earth continuity resistance between the main earthing point and any other position in the completed installation must comply with IEE Tables 41.2, 41.3, 41.4 and 41.5.

A range of glands and locknuts is available for entering the cables into any standard boxes or casings designed to take steel conduit. The glands, which are

TABLE 14.1 Minimum Spacing of Fixings for MI Copper-sheathed cables.
These Spacings are for Cables in Accessible Positions

Overall diameter of cable (mm) of mineral-insulated copper-sheathed-cables	Horizontal (mm)	Vertical (mm)
Not exceeding 9	600	1800
Exceeding 9 and not exceeding 15	900	1200
Exceeding 15 and not exceeding 20	1500	2000

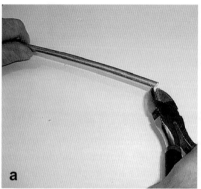
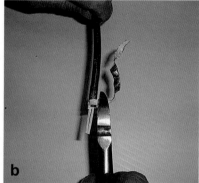

(a) Preparing cable end. Strip cable end by gripping the edge of the sheath between the jaws of side-cutting nippers and twist the cable off in stages, keeping the nippers at about the angle shown. (b) Then proceed in a series of short rips, pulling off the sheath in a spiral.

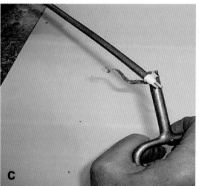
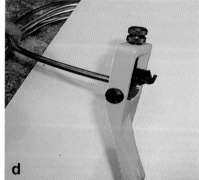

(c) An alternative method of stripping sheath to expose long conductors. A stripping rod, which can be easily made from a piece of mild steel is used in a similar manner to a tin opener. (d) The MI cable joistripper being used to start the cut to remove the cable sheath.

FIGURE 14.4   Mineral insulated cable – stripping the sheath.

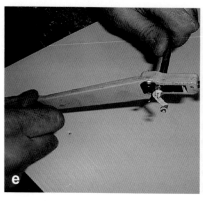
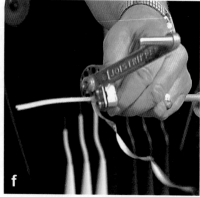

(e) And seen here removing the sheath. (f) Large rotary stripping tool (all Wrexham Mineral Cables).

FIGURE 14.4 cont'd. Mineral insulated cable – stripping the sheath.

slipped onto the cables before the cable ends are sealed, firmly anchor the cables and provide an efficient earth bonding system.

In some instances, it may not be possible to ensure bonding via the gland, e.g. when fixed into a plastic box. In these instances, a seal is available which incorporates an additional earth tail wire.

Regulations on Sealing

The ends of MI metal-sheathed cables must be sealed to prevent the entry of moisture and to separate and insulate the conductors.

The sealing materials shall have adequate insulating and moisture-proof properties, and shall retain these properties throughout the range of temperatures to which the cable is subjected in service. The manufacturers provide a plastic compound for use on the standard cold screw-on pot type seal.

Methods of stripping and sealing are given below. The tools required include hacksaw, side cutting pliers, screwdriver, special ringing tools and pot wrench.

14.3 PREPARATION OF CABLE END

To prepare the ends of the cable prior to sealing, cut the cable to length with a hacksaw. The sheath is then scored with a ringing tool to enable a clean end to be made when the sheath is removed. Tighten the nut of the ringing tool so that the wheels JUST grip the sheath and then give the nut a further quarter to half a turn. By rotating the tool through 360° or more, the sheath will be ringed. If the ring is made too deep it will be found difficult to break into it when stripping; if too shallow the sheath will be bell-mouthed and the gland and seal parts will not readily fit onto the sheath. If there is any roughness left around the end of the sheath from the ringing tool, remove it by lightly running the pipe grip part of a pair of pliers over it. If the length to be stripped is very long, defer ringing until stripping is within 50–75mm of the sealing point.

After ringing at the sealing point, strip the sheath to expose the conductors. Use side-cutting pliers to start the 'rip'. To do this, grip the edge of the sheath between the jaws of the pliers and twist the wrist clockwise, then take a new grip and rotate through a small angle. Continue this motion in a series of short 'rips' keeping the nippers at about 45° to the line of the cable, removing the sheath spirally. When about to break into the ring, bring the nippers to right angles with the cable. Finish off with point of nippers held parallel to the cable.

An alternative method of stripping, often employed for long tails, is to use an easily constructed stripping rod, as illustrated. This can easily be made from a piece of mild steel rod, about 10 mm in diameter, the end slot being made by a hacksaw. Start the 'rip' with pliers then pick up the tag in the slot at the end of

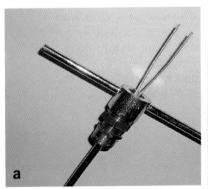

(a) A quick and accurate method of fitting the pot is by the use of a pot wrench. (b) Alternatively, a wrench can be used to fit the pot.

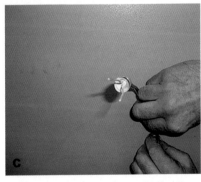
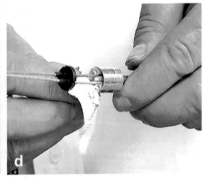

(c) Examine the inside of the pot for cleanliness and metallic hairs prior to filling with compound. (d) Overfilling the pot with compound. Use the plastic wrapping to prevent fingers coming into contact with the compound so as to ensure cleanliness of the seal.

FIGURE 14.5   Sealing the cable end.

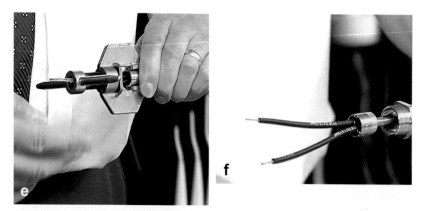

(e) Securing the stub cap in position using a crimping tool which makes three indent crimps. Finally the pot is crimped using the crimping tool. (f) The termination is completed by sliding on the insulating sleeves (all Wrexham Mineral Cables).

FIGURE 14.5   cont'd.   Sealing the cable end.

the rod and twist it, at the same time taking it round the cable; break into the ring and finish as with the nipper method.

For light duty cables up to 4L1.5 in size the Joistripper tool is very efficient, it is quick and easy to use, and will take off more sheath than any other tool of its type, and is available from the manufacturers of MI cables, and their suppliers. For other cables, large or small rotary strippers can be used, these are also obtainable from the cable manufacturers.

14.4 SEALING CABLE ENDS

The standard screw-on seal consists of a brass pot that is anchored to the cable sheath by means of a self-tapping thread. The pot is then filled with a sealing compound and the mouth of the pot is closed by crimping home a stub cap or disc/sleeve assembly. The components necessary are determined by the conductor temperature likely to be encountered. They are as follows:

- 80 to 105 °C Grey sealing compound, stub cap with PVC stub sleeving or fabric disc with headed PVC sleeving, for standard seals.
- 20 to 60 °C Grey sealing compound, red/pink polypropylene disc with headed PVC sleeving, for increased safety seals.

Having ringed and stripped the sheath, slip the gland parts, if any, onto the cable. To complete the screw-on seal, see that the conductors are clean and dry, engage the sealing pot square and finger tight on the sheath end. Then tighten the pot with pliers or grips until the end of the cable sheath is in level with the shoulder at the base of the pot. In general the cable should not project into the pot but a 1 or 2mm projection is required for certain 250 °C and increased safety seals. Alternatively the pot wrench can be used in conjunction with the gland body.

If the pot is difficult to screw on, moisten the sheath with an oil damped rag. To avoid slackness do not reverse the action. Examine the inside of the pot for cleanliness and metallic hairs, using a torch if the light is poor. Test the pot for fit inside the gland. Set the conductors to register with the holes in the cap. Slip the cap and sleeving into position to test for fit, and then withdraw slightly. Press compound into the pot until it is packed tight. The entry of the compound is effected by feeding in from one side of the pot only to prevent trapping air. To ensure internal cleanliness of the seal, use the plastic wrapping to prevent fingers from coming into contact with the compound (Fig. 14.5d).

Next slide the stub cap over the conductors and press into the recess in the pot. Finally, the pot must be crimped using a crimping tool and the termination completed by sliding insulated sleeves of the required length onto the conductors. New types of seal are becoming available with the sealing compound supplied as an integral part of the seal. These seals are easier to fit and information on their use may be obtained from cable manufacturers.

14.5 CURRENT RATINGS OF CABLES

Owing to the heat-resisting properties of MI cables and to the fact that the magnesia insulation is a good conductor of heat, the current ratings of these cables are higher than those of PVC or even PI cables.

Multi-core cables are not made larger than $25mm^2$, and therefore when heavier currents need to be carried it is necessary to use two or more single-core cables which are made in sizes up to $240mm^2$. Where single-core cables are run together their disposition should be arranged as shown in IEE Table 4A2. The current-carrying capacity of large single-core cables depends considerably upon their disposition.

IEE Tables 4G1A–4G2A are for copper conductor MI cables. When these cables are run under conditions where they are not exposed to touch, they are rated to run at a comparatively high temperature and the current rating is considerably more than cables which are exposed to touch, or are covered with PVC sheathing. For example, a 150-mm^2 single-core cable is rated to carry 388A if exposed to touch, but if not exposed to touch the same cable is rated to carry 485A.

When an installation is designed to carry these higher currents, due regard must be paid to voltage drop, and also to the fact that the high temperature which is permissible in these cables might be transmitted to switchgear, and which might be affected by the conducted heat from the cable.

14.6 SOME PRACTICAL HINTS

These cables are supplied in coils, and every effort should be made to ensure that the coils retain their circular shape. They are frequently thrown off the delivery lorry and the impact flattens and hardens them. Before despatch the manufacturers anneal the cables so they are in a pliable state, but during transit

FIGURE 14.6 Emergency stop and fire alarm are grouped together in this industrial installation. MI cable in use to connect the break glass fire alarm control.

and subsequent handling manipulation in excess of the manufacturers' recommendations will harden the cable and could cause sheath fracture.

To measure the cable it should not be run out and recoiled as this tends to harden the cable. The best way is to measure the mean diameter of the coil and multiply by 3.14 which will give the approximate length of each turn in the coil.

Kinks or bends in the cable can best be removed by the use of a cable straightener. This is a device with pressure rollers that can be run backwards and forwards over the cable until the kinks are smoothed out.

The magnesium oxide insulation used in the cable has an affinity for moisture. There is, therefore, a need for temporary sealing during storage.

After sealing, an insulation test between conductors and to earth should be carried out, and this test should be repeated not less than 24 h later. The second reading should have risen, and be at least 100MV with a 500V insulation tester.

As the conductors cannot be identified during the manufacturing process it is necessary to identify them after making off the seals. This can be done by fitting coloured sleeves or numbered markers onto the core. Correct identification can be checked by the use of a continuity tester.

14.7 INDUCTIVE LOADS

Switching of inductive loads can cause high voltage surges on 230V and 400V circuits, and these surges could cause damage to MI cables. Protection from these surges can be achieved by the use of inexpensive surge suppressors. The manufacturers of MI cables will be pleased to give advice on this matter.

Chapter 15

Luminaires, Switches, Socket Outlets and Data Circuits

The final stage of electrical installation work is the fixing of accessories, such as ceiling roses, holders, switches, socket outlets, luminaires and, with many office and commercial installations, connecting data circuits. This work requires experience and a thorough knowledge of the regulations which are applicable, because danger from shock frequently results from the use of incorrect accessories or due to accessories being wrongly connected.

IEE Regulation 133 lays down the requirements for the selection of equipment.

15.1 CEILING ROSES

Ceiling roses may be of the two-plate pattern and must also have an earth terminal (Fig. 15.2). The three-plate type is used to enable the feed to be looped at the ceiling rose rather than to use an extra cable which would be needed to loop it at the switch.

For PVC-sheathed wiring it is possible to eliminate the need for joint boxes if three-plate ceiling roses are employed (see Chapter 13). No ceiling rose may be used on a circuit having a voltage normally exceeding 250V. Not more than two flexible cords may be connected to any one ceiling rose unless the latter is specially designed for multiple pendants.

Special three- and four-pin removable fittings rated at 2A or 6A may be obtained and these can be installed where lighting fittings need to be removed or rearranged. The ability to remove lighting easily can assist in carrying out maintenance. These connectors may not be used for the connection of any other equipment [IEE Regulation 559.6.1.4].

For the conduit system of wiring it is usual to fit ceiling roses which screw directly on to a standard conduit box, the box being fitted with an earth terminal.

15.2 LUMINAIRES AND LAMPHOLDERS

Every luminaire or group of luminaires must be controlled by a switch or a socket outlet and plug, placed in a readily accessible position.

FIGURE 15.1   A range of fittings for final circuits including socket outlets, switches, an accessory switch with a neon indicator light and a telephone socket.

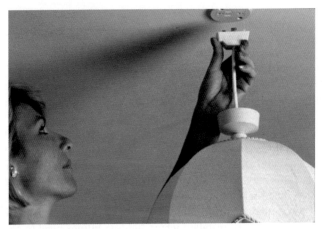

FIGURE 15.2   A three-pin connector rated at 2A designed to enable lighting to be easily removed and refitted (Ashley and Rock Ltd).

In damp situations, every luminaire shall be of the waterproof type, and in situations where there is likely to be flammable or explosive dust, vapour or gas, the luminaires must be of the flameproof type.

A number of requirements apply to luminaires and lampholders and these are covered in IEE Regulation 559. Insulated lampholders should be used wherever possible. Lampholders fitted with switches must be controlled by a fixed switch or socket outlet in the same room.

(a) The lighting fitting is to be secured on steel trunking which has already been installed. The body of the fitting is supported whilst the cables are threaded through the brass bush. (b) After securing the body to the conduit, the feeds are cut to length and connected up.

(c) The fitting is assembled taking care not to trap any of the conductors. Where the option exists, the use of 'twin and earth' cable can make assembly easier, as there is less risk of trapping. (d) The tube is fitted.

(e) The diffuser clipped in place. Some designs allow clipping from either side of the diffuser which assists with assembly and subsequent lamp changing. (f) The completed fitting in position.

FIGURE 15.3 Installing a fluorescent lighting fitting.

The outer screwed contact of Edison screw-type lampholders must always be connected to the neutral of the supply. Small Edison screw lampholders must have a protective device not exceeding 6A, but the larger sizes may have a protective device not exceeding 16A.

No lampholder may be used on circuits exceeding 250V (IEE Regulation 559.6.1.2), and all metal lampholders must have an earth terminal. In bathrooms, and other positions where there are stone floors or exposed extraneous conductive parts, lampholders should be fitted with insulated skirts to prevent inadvertent contact with live pins when a lamp is being removed or replaced.

15.3 FLEXIBLE CORDS

The definition of a 'flexible cord' is 'A flexible cable in which the cross sectional area of each conductor does not exceed $4mm^2$'. Larger flexible conductors are known as 'flexible cables'. Flexible cords, if not properly installed and maintained, can become a cause of fire and shock. They must not be used for fixed wiring.

Flexible cords must not be fixed where exposed to dampness or immediately below water pipes. They should be open to view throughout their entire length, except where passing through a ceiling when they must be protected with a properly bushed non-flammable tube. Flexible cords must never be fixed by means of insulated staples.

FIGURE 15.4 Suspended ceilings are regularly encountered in commercial premises. The matrix is designed to accept air conditioning, fire detection and lighting fittings, as seen in this view. The luminaires are Dextra compact fluorescent fittings with reflectors which give indirect as well as direct illumination (W.T. Parker Ltd).

Where flexible cords support luminaires the maximum weight which may be supported is as follows:

$0.5mm^2$	2kg
$0.75mm^2$	3kg
$1.0mm^2$	5kg

In kitchens and utility rooms, and in rooms with a fixed bath, flexible cords shall be of the PVC sheathed or an equally waterproof type.

When three-core flexible cords are used for fixed or portable fittings that have to be earthed, the colour of the cores shall be *brown* (connected to phase side), *blue* (connected to neutral or return), and *green/yellow* (connected to earth). When four-core flexible cords are used for three-phase appliances, the colours of the cores shall be *brown*, *black* and *grey* for the phases, *blue* for neutral, with *green/yellow* for the protective conductor.

Connections between flexible cords and cables shall be effected with an insulated connector, and this connector must be enclosed in a box or in part of a luminaire. If an extension of a flexible cord is made with a flexible cord connector consisting of pins and sockets, the *sockets* must be fed from the supply, so that the exposed pins are not alive when disconnected from the sockets.

Where the temperature of the luminaire is likely to exceed 60 °C, special heat-resisting flexible cords should be used for all tungsten luminaires, including pendants and enclosed type luminaires, the flexible cord should be insulated with butyl rubber or silicone rubber. Ordinary PVC insulated cords are not likely to stand up to the heat given off by tungsten lamps. Flexible cords feeding electric heaters must also have heat-proof insulation such as butyl or silicone rubber.

Where extra high temperatures are likely to be encountered it is advisable to consult a cable manufacturer before deciding on the type of flexible cord to be used.

Flexible cords used in workshops and other places subjected to risk of mechanical damage shall be PVC sheathed or armoured. All flexible cords used for portable appliances such as portable handlamps, floor and table lamps shall be of the sheathed circular type.

All flexible cords should be frequently inspected, especially at the point where they enter lampholders and other accessories, and renewed if found to be unsatisfactory.

15.4 SOCKET OUTLETS AND PLUGS

The 13A socket outlet with fused plug made to BS 1362 and BS 1363 is in general use for domestic and office premises (Figs 15.5 and 15.6). The 13A socket outlet is also extensively used in industrial premises. Socket outlets to BS 196 are also used for circuits not exceeding 250V, and are made in ratings of 5A, 15A and 30A. Other industrial type socket outlets are covered by BS EN 60309, and these include single-phase and three-phase with ratings up to 125A.

FIGURE 15.5   A range of colourful 13-A sockets for use in dwellings or offices.

Details of the ratings and circuiting of these various types of socket outlets are given in Chapter 5.

The Low Voltage Electrical Equipment (Safety) Regulations 1989 require equipment to be safe. This implies that any part intended to be electrified is not to be capable of being touched with a finger, and this includes a child's finger. Thus the live pins of plugs should be partly shrouded so that when the plug is in

FIGURE 15.6   The 13A plug, for attachment permanently to the appliance for which it will be used, can carry a fuse of suitable rating up to 13A to provide individual protection for that appliance. The plugs shown are of two types. That showing the interior with fuse and cable grip arrangement is available to be fitted to the appliance. The other type (centre) is moulded to the flexible cable in the factory and cannot be subsequently removed. Both types have sleeved pins as required by BS 1363.

the process of being inserted even the smallest finger cannot make contact with live metal. BS 1363, Clause 4-2-2 requires that socket outlets shall be provided by a screen which automatically covers the live contacts when the plug is withdrawn.

When installing socket outlets the cables must be connected to the correct terminals, which are:

brown wire (phase or outer conductor) to terminal marked L,
blue wire (neutral or middle conductor) to terminal marked N and
yellow/green earth wire to terminal marked E.

Flexible cords connected to plugs shall be brown (phase), blue (neutral) and yellow/green (earth). If wrong connections are made to socket outlets it may be possible for a person to receive a shock from an appliance, even when it is switched off (Fig. 15.7).

Socket outlet adaptors which enable two or more appliances to be connected to a single socket should contain fuses to prevent the socket outlet from becoming overloaded.

Socket outlets installed in old people's homes and in domestic premises, likely to be occupied by old or disabled people, should be installed at not less than 1 m from floor level.

15.5 SWITCHES

There are various types of switches available, the most common being the 6A switch which is used to control lights. There is also the 16A switch for circuits carrying heavier currents. The correct method of mounting switches

FIGURE 15.7   A five-pin three-phase + neutral + earth industrial shuttered socket to BS EN 60309-2.

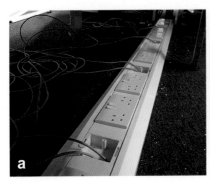

(a) This multi-compartment trunking has the power socket installation complete and the data wiring (run in a separate compartment) is ready for connection. (b) A typical data socket with two separate outlets mounted side by side. The terminals are pre-marked with the colour code for the connections.

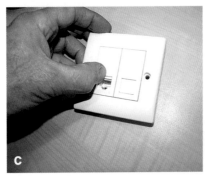

(c) Front view of the twin socket with one of the shutters open to view the contacts inside. (d) Cores must be numbered during installation to ensure connection to the correct outlet.

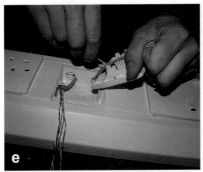

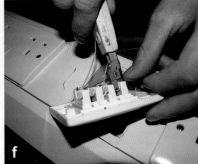

(e) After stripping the outer sheath, the cores are separated and the data cable is secured to the cable grip using the cable tie provided. (f) Individual conductors are pressed in to the terminal using a Cronin tool.

FIGURE 15.8 Installing and connecting data circuits.

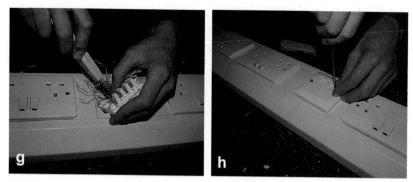

(g) This tool simultaneously clips off the surplus conductor. It is important to install the coloured cores into the correct terminal on the socket. (h) After completing the connection of all 16 cores, the conductors are carefully disposed and the socket screwed to the outlet box.

(i) A close-up of the Cronin tool with integral clipper (all M.W. Cripwell Ltd).

FIGURE 15.8 cont'd. Installing and connecting data circuits.

for the various wiring systems is dealt with in the sections which cover these systems.

All single-pole switches shall be fitted in the same conductor throughout the installation, which shall be the phase conductor of the supply.

In damp situations, every switch shall be of the waterproof type with suitable screwed entries or glands to prevent moisture entering the switch or socket. To prevent condensed moisture from collecting inside a watertight switchbox, a very small hole should be drilled in the lowest part of the box to enable the moisture to drain away. Flameproof switches must be fitted in all positions exposed to flammable or explosive dust, vapour or gas.

15.6 DATA CIRCUITS

Almost every commercial and industrial installation makes considerable use of data circuits. In general, a data or computer room is set aside for the processors and communication equipment and radial data and telephone circuits are run out to various parts of the office, commercial or production areas. The most common form of data circuit is wired in eight-wire (four twisted pairs) communication cable and this is often run in one compartment of multi-compartment trunking or cable basket where it can be segregated from power and lighting circuits.

Cable is available to a range of specifications depending on whether voice, data or video is to be transmitted. There is much to be said for wiring an installation in such a way that the various applications can all be made in the future. Screening is needed for some uses. Cable is supplied with the conductors in twisted pairs and can be of type UTP (unscreened), FTP (foil screened) or S/FTP (foil and braid screened). Most installations use unscreened (UTP) cable.

Cable connections are made without the need to strip the insulation from the individual cores. The Cronin press tool which is used is specifically designed for the purpose and the terminal slot in the socket is of such a dimension that metallic contact is made as the cores are inserted into position. The Cronin tool incorporates a cutter which simultaneously cuts off the surplus conductor.

Chapter 16

Inspection and Testing

16.1 INTRODUCTION

The Purpose of Inspection, Testing and Certification or Reporting

The fundamental reason for inspecting and testing an electrical installation is to determine whether new installation work is safe to be put into service, or an existing installation is safe to remain in service until the next inspection is due.

Required Competence to Undertake Electrical Inspection and Testing

The inspector carrying out the inspection and testing of any electrical installation must have a sound knowledge and experience relevant to the nature of the installation being inspected and tested, and to the technical standards. The inspector must also be fully versed in the inspection and testing procedures and employ suitable testing equipment during the inspection and testing process.

Safety

Electrical testing involves some degree of hazard and before the commencement of any tests, the inspector must take steps to ensure that they work in a safe manner and also consider the safety of others when the test takes place. The safety procedures detailed in health and safety Executive Guidance Note GS38, electrical test equipment for use by electricians, should be observed.

Before any testing takes place the tester should ensure that the meter is within the calibration date and has been checked for ongoing accuracy before use. If the test instrument is not within the calibration date the results obtained will be classified as invalid. The test meter has to be proved accurate before commencement of any testing takes place.

General information and guidance on Safety procedures can be found in IEE Guidance Note 3, Section 1. Inspection and testing of electrical installations are dealt with in Part 6 of the IEE Regulations, Chapters 61, Initial Verification, and 62, Periodic Inspection and Testing. Chapter 63 of the IEE Regulations lays out the recommendations when certifying and reporting on an Electrical Installation.

In this chapter, two aspects will be considered. Firstly, Initial Verification of the installation, followed by Periodic Inspection and Testing. The first of these is covered within IEE Chapter 61. This section covers and recommends procedures in the inspection of new electrical installations.

16.2 INITIAL VERIFICATION

General Procedure

Initial verification, in the context of the IEE Regulations, is covered by Regulation 610, which is intended to confirm that the installation complies with the designer's intentions and has been constructed, inspected and tested in accordance with BS 7671, the IEE Regulations.

Before the commencement of initial inspection and testing the designer, or the person responsible for the design, must make available the results of the assessment of general characteristics required by IEE Sections 311–313, together with the information required by Regulation 514.9. IEE Regulation 610 Inspection and, where appropriate, testing should be carried out and recorded on suitable schedules progressively throughout the different stages of erection and before the installation is certified and put into service.

The results of the different stages of testing must be compared with the design calculations, so as to determine that the correct installation procedures have taken place and that the design, on completion, will comply with the appropriate mandatory, statutory regulations, British Standards and building regulations.

Inspection

IEE Regulation 611 requires the checking of a number of items in the installation and that where necessary this should be done during erection. These include:

- electrical connections
- identification of conductors
- safe routing of cables
- conductors are selected in accordance with the design
- that single-pole devices are connected in the phase conductor
- correct connection of sockets, accessories and equipment
- presence of fire barriers
- appropriate insulation of conductors
- presence of protective conductors
- appropriate isolators and switches
- methods of protection against electric shock
- prevention of mutual detrimental influence
- undervoltage protection

- danger notices and labelling of circuits, fuses etc.
- access to switchgear is adequate.

Testing

IEE Regulations 612.1–612.14 detail the standard methods of testing required. The tests should be as follows, and should be carried out in the sequence indicated:

1. continuity of protective conductors
2. continuity of final circuit ring conductors
3. insulation resistance
4. insulation of site-built assemblies
5. protection by separation of circuits
6. protection by barriers or enclosures
7. insulation of non-conducting floors and walls
8. polarity
9. earth electrode resistance
10. earth fault loop impedance
11. prospective fault current
12. functional tests including the operation of residual current devices (RCDs).

The methods and recommendations of carrying out the initial inspection and testing of electrical installations are detailed in IEE Regulations Chapter 61 along with guidance on initial inspection and testing.

Most installations will be covered by the test methods described in the IEE Guidance Note 3 Section 2 and the IEE Regulations state the preferred testing methods to be used. If installation concerned does not come into the remit of these publications, guidance on inspection and testing methods must be sought from one of the companies' qualifying supervisors before commencement of any initial inspection and testing.

The Health and Safety Executive has issued a guide on Electrical Testing, HS (G) 13, which gives advice on precautions which should be taken when testing live installations. The guide mentions that many accidents occur when making these tests. It recommends that bare ends of test probes should not exceed 2–3mm of bare metal, and that metal lampholders should never be used for test lamps.

Some further advice, based upon practical experience, is given here to supplement the advice contained in the Regulations.

Continuity Tests

The requirements for continuity testing are covered in Section 612.2 of the IEE Regulations.

Test Instrument

The test instrument to be used for continuity testing is an ohmmeter having a low ohms range, or an insulation and continuity test instrument set to the continuity range. Continuity test readings of less than 1 ohm are common. Therefore, the resistance of the test leads is important, and should not be included in any recorded test results. If the test instrument being used does not have provision for correcting the resistance of the test lead, it will be necessary to measure the resistance of the leads when connected together, and the measured value to be subtracted from all the test results.

Continuity of Circuit Protective Conductors

There are two main methods of continuity testing, and these are described below.

Method 1

The line conductor is required to be connected to the protective conductor within the distribution board, this commonly achieved by using a bridging strap connected between the line conductor and the relevant earth connection related to the particular circuit. Then with a continuity tester test between the line and earth terminals at each point in the circuit, the measurement at the circuit's extremity should be recorded, which will be the value of $(R_1 + R_2)$ for the circuit under test.

If the instrument does not include an 'auto-null' facility, or this is not used, the resistance of the test leads should be measured and deducted from the final resistance reading of the circuit under test.

FIGURE 16.1 The test for continuity involves using a bridging strap connected between the line conductor and the relevant earth connection related to the particular circuit, here conveniently applied at a socket outlet.

Method 2

This is commonly known as the long lead test. This test requires the use of a long piece of cable connected to the main earth terminal at source. The other end of the cable is then connected to one side of the continuity meter. The remaining terminal on the continuity is then connected onto the protective conductor at various points on the circuit under test, such as luminaries, switches, spur outlets etc. The results obtained, after the deduction of the resistance of the long lead, are recorded on test certificates under the column of R_2.

This particular test may also be utilised to prove the continuity of the main equipotential and extraneous earthing conductors, although if the earthing cables are visible and if it is possible to trace the cable from the origin to the destination, this test may not be required.

Continuity of Ring Final Conductors

There are two methods for testing a ring circuit, which are detailed below.

Method 1

The first test is to prove the continuity of each conductor and to prove that the conductor continues throughout the circuit. This is achieved by testing the continuity of ring final circuit conductors, a digital ohmmeter or multimeter set to 'ohms' range should be used. The ends of the ring circuit conductors are separated and the resistance values noted for each of the live conductors and for the protective conductor. The ring circuit is then reconnected and a further resistance measurement taken for each conductor between the distribution board and the appropriate pin of the outlet nearest to the mid-point of the ring. The value obtained should be approximately one quarter of the value of the first reading for each conductor. The test lead needed to carry out the second part of this test will be quite long, and it will be necessary to determine its resistance and deduct the figure from the readings obtained to obtain a valid result.

Method 2

An alternative method of testing a ring circuit avoids the use of a long test lead. It is initially necessary to determine which ends are which for the installed ring circuit. This is done by shorting across the phase and neutral conductors of the first or last socket outlet on the ring, and applying an ohmmeter to the cable ends at the distribution board (see Fig. 16.2a). If the readings of the test meter are different in position A than in position B the pairs are matched correctly and the test may be continued. If the readings are the same in position A and in position B, the short and long sides of the ring are linked, and the wrong pairs have been selected, therefore the test is unacceptable.

The next step is to remove the short circuit from the first or last socket outlet on the ring. Then short together the live conductor of one of the pairs of cables and the neutral conductor of the other; also short together the remaining pair of cables (see Fig. 16.2b).

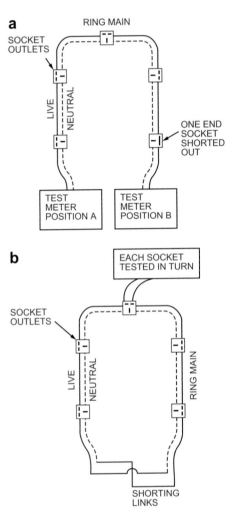

FIGURE 16.2 Testing the continuity of a ring circuit as described in the text: (a) indicates the end socket shorted out for the initial test to identify the individual cables; (b) shows part two of the test, applying the ohmmeter to each socket in turn and comparing the resistance readings. All the resistances must be identical to show that continuity is proved.

The test instrument is then connected to each socket outlet on the ring in turn. The resistance reading in each position should be identical, and if it is, the continuity is proved. If one of the readings is different, the socket outlet either is connected as a spur to the ring circuit or is a socket outlet on a different ring.

Special instruments are available for checking the resistance of the CPC or of metal conduit and trunking where it is used as part of the protective conductor. The instrument operates by applying a current at extra-low voltage

TABLE 16.1 Minimum Values of Insulation Resistance for Standard Circuits

Circuit nominal voltage (V)	Test voltage (V d.c.)	Minimum insulation resistance (MΩ)
SELV and PELV	250	≥0.5
Up to and including 500V with the exception of SELV and PELV, but including FELV	500	≥1

to the section of conduit or trunking connected and gives a reading of the continuity in ohms. The characteristics of such an instrument are included in IEE Regulation 713-02-01.

Insulation Resistance

Insulation tests should be made with an insulation resistance tester, with a scale reading in ohms. The voltage of the instrument should be as detailed within Tables 16.1 and 16.2.

Suitable instruments for making these tests are shown in Figs 16.3–16.5.

The main test should be made before the luminaires and lamps are installed, but with all fuses inserted, all switches on, and the conductors of both poles connected together, and with the supply switched off. This test will be between all conductors bunched, and earth. The result of the test should be not less than 1.0MΩ. Before the test, particular attention should be given to the presence of electronic devices connected to the installation, and such devices should be isolated so that they are not damaged by the test voltage.

Another test is between phase and neutral conductors, with all lamps removed, and all switches in the 'on' position. This test shall produce a reading

TABLE 16.2 Minimum Values of Insulation and Resistance for SELV, PELV and Circuits Above 500V

Circuit nominal voltage (V)	Test voltage (V d.c.)	Minimum insulation resistance (MΩ)
SELV and PELV	250	≥ 0.5
Above 500V	1000	≥ 1

Note: Test voltages to be as follows:

250V d.c. for extra-low voltage circuits

500V d.c. for low-voltage circuits up to 500V, and

1000 V d.c. for low-voltage circuits between 500V and 1000V.

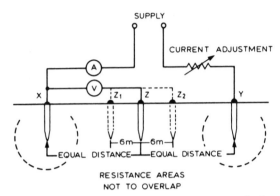

FIGURE 16.3 Measurement of earth electrode resistance. X – earth electrode under test, disconnected from all other sources of supply; Y – auxiliary earth electrode; Z – second auxiliary earth electrode; Z_1 – alternative position of Z for check measurement; Z_2 – further alternative position of Z for check measurement. If the tests are made at power frequency the source of the current used for the test shall be isolated from the mains supply (e.g. by a double-wound transformer), and in any event the earth electrode X under test shall be disconnected from all sources of supply other than that used for testing.

of not less than $1.0\text{M}\Omega$. If a reading lower than $1.0\text{M}\Omega$ is obtained then steps must be taken to trace and rectify the fault.

Where surge protective devices, electronic equipment or other devices such as RCDs are present, these are likely to influence the results of the test and may suffer damage from the test voltage. Such equipment must be disconnected before carrying out the insulation resistance test.

If it is not reasonably practicable to disconnect electronic related equipment, the recommended test voltage for the type of circuit may be reduced to 250V d.c. but the insulation resistance must be at least $1\text{M}\Omega$.

FIGURE 16.4 A 250/500/1000V insulation and continuity tester, with digital display.

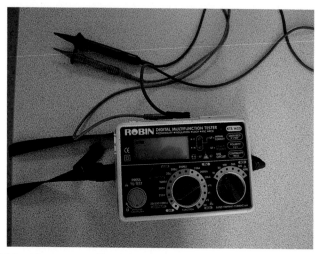

FIGURE 16.5   A battery-operated insulation resistance test instrument operates at 250, 500 or 1000V.

Polarity

The testing for polarity is the same as the check previously carried out earlier in the test sequence. Ring final circuits require a visual check although, as with the radial circuits, the test for ring main circuits was previously carried out when testing the continuity of the circuit.

IEE Regulation 612.6 requires that every fuse and single-pole control and protective device are connected in the line conductor only. It also requires a check that E14 and E27 lampholders, not to BS EN 60238, have the outer or screwed contacts connected to the neutral conductor; but this does not apply to new installations, as new lampholders should be of BS EN 60238 type.

Earth Electrode Resistance

The test should be carried out with a digital earth tester. An alternating current is passed between points X and Y and an additional earth spike Z is placed successively at points Z_1, Z_2 etc. Voltage drops between X and Z, and Z and Y are obtained for successive positions of Z and the earth electrode resistance is calculated and checked from the voltage drop and current flowing.

Earth Fault Loop Impedance

Tests for earth fault loop impedance should be made with an instrument such as that shown in Figs 16.6 and 16.7.

The object of this test (Fig. 16.8) is to ensure that the phase earth loop impedance of the circuit is appropriate to the rating and type of protective device as specified by the IEE Regulations and thus ensure that the circuit will

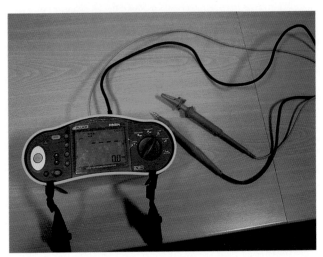

FIGURE 16.6 An instrument suitable for use as a digital loop tester is shown here. As with a number of instruments on the market, it will operate in several modes, easing the work of the tester.

disconnect within the correct time. If a fault did not result in the fuse or circuit breaker disconnecting in the correct time, a very dangerous state of affairs could exist, and it is important that this test be made and acted upon.

Testing Residual Current Circuit Breakers

The operation and use of residual current circuit breakers were described in Chapter 2. Test instruments can be obtained which are designed to carry out tests of RCDs and the instrument is connected to the load side of the device, the loads themselves being disconnected. The test instrument simulates a fault so

FIGURE 16.7 The same instrument in use for a continuity test.

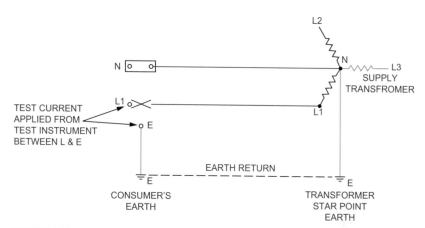

FIGURE 16.8 Earth fault loop impedance test measures the impedance in the line-earth loop which comprises the following parts: the circuit protective conductor; the consumer's earthing terminal and earthing conductor; the earth return path through the general mass of earth; the supply transformer earth; the neutral point of the supply transformer and winding; the phase conductor.

that a residual current flows, and then measures the response time of the RCD, generally displaying the result in milliseconds. RCDs incorporate an integral test button and the effectiveness of this should also be tested (Fig. 16.9).

The installation of voltage operated earth leakage circuit breakers is not now permitted by the IEE Regulations. However, their use may be encountered in existing installations, and details of a test method suitable for them are given in Fig. 16.10. Voltage operated devices have a number of disadvantages and if any

FIGURE 16.9 This digital instrument is capable of testing RCDs. Devices with ratings from 30mA to 1000mA may be tested.

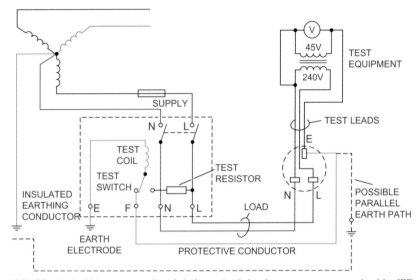

FIGURE 16.10   Voltage operated earth leakage circuit breakers are not now permitted by IEE Regulations. However, their use may be encountered in existing installations, and testing of them may be carried out using this circuit. A test voltage not exceeding 50V a.c. obtained from a double-wound transformer of at least 750VA is connected as shown. If satisfactory the circuit breaker will trip instantaneously.

doubt exists as to their performance, they should be replaced by residual current circuit breakers.

Other tests included in the IEE Regulations are phase rotation tests and nominal voltage tests. These tests clarify that the voltages present are within the required parameters for the type of installation and as recommended within the Electricity Safety, Quality and Continuity Regulations 2002 (ESQCR).

Alterations and Additions to an Installation

The relevant requirements of Section 633 of the IEE Regulations apply to alterations and additions to installations. It shall be verified that every alteration or addition complies with the regulations and does not impair the safety of an existing installation.

16.3 PERIODIC INSPECTION AND TESTING

Purpose of Periodic Inspection

The main purpose of periodic inspection and testing is to detect, so far as is reasonably practicable, and to report on, any factors impairing the safety of an electrical installation.

The aspects to be covered are stated in IEE Regulation 621.2 and include the following:

a. Safety of persons and livestock against the effects of electric shock and burns.
b. Protection against damage to property by fire and heat arising from an installation defect.
c. Confirmation that the installation is not damaged or deteriorated so as to impair safety.
d. Identification of non-compliances with BS 7671 or installation defects which may give rise to danger.

Necessity for Periodic Inspection and Testing

Periodic inspection and testing is necessary because all electrical installations deteriorate due to a number of factors such as damage, wear and tear, corrosion, excessive electrical loading, ageing and environmental influences.

Required Information

It is essential that the inspector knows the extent of the installation to be inspected and any criteria regarding the limit of the inspection, this should be recorded.

Enquiries need to be made with regards to the provision of diagrams, design criteria, electricity supply and earthing arrangements. These will normally be obtained from the person in charge of the installation. Diagrams, charts or tables should be available to indicate the type and composition of circuits, identification of protective devices for shock protection, isolation and switching and a description of the method used for 'fault protection' before the commencement of any periodic inspection and testing takes place.

If the required information is not available, then the person carrying out the inspection should make their own assessment of all perimeters of the electrical installation. In this case, on completion of the inspection, an as-fitted drawing of the electrical installation should accompany the Periodic Inspection and Test results.

These records should be retained for further works and inspections, so as to identify any alterations or additions that may occur after the undertaken inspection. If the building facilities manager keeps a building log book the information from the test results may have to be recorded and copied into the appropriate sections of the log book or building manual (Fig. 16.11).

Schedule of Inspection and Testing

No electrical testing should be performed on an installation that does not comply with the current legislation with regards to main, equipotential and supplementary bonding. Testing with the bonding being absent could inadvertently cause any extraneous metal parts or metal parts directly related to the electrical installation to become live. Undertaking the tests in these conditions

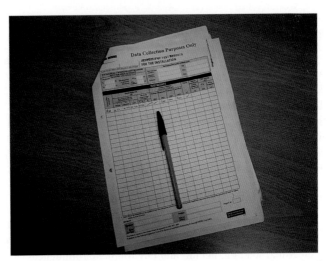

FIGURE 16.11 All readings must be recorded at the time of testing and pre-printed sheets are available for the standard tests.

may contravene sections of the Electricity at Work Regulations 1989 and the Health and Safety at Work Act 1974, Section 6.

Frequency of Inspection

The frequency depends upon the general condition of the electrical installation. If the installation tested is not to standard, then it might be prudent for safety reasons to set the next inspection date for a period less than that indicated in IEE Guidance Note 3, Table 3.3. IEE Table 3.2 which recommends initial frequencies of inspection has been altered to coincide with the recommendations of the report and the IEE Regulations.

Installations That May Require Periodic Visual Inspection

If an installation is maintained under a planned maintenance management system which incorporates monitoring and is supervised by a suitably qualified electrical engineer then a formal periodic inspection and test certificate may not be required.

A visual inspection in line with IEE Guidance Note 3, Section 3.5 and page 4 of the NICEIC Periodic Inspection form should record the basic information including:

a. The characteristics of the main device,
b. The earthing arrangements,
c. The size and continuity of equipotential and supplementary bonding conductors,
d. Functional test of RCDs,

e. Functional test of circuit breakers, isolators and switching devices and

f. Earth fault loop impedance values should be sampled and cross-referenced with the existing/previous test results for comparison.

The records may be kept on paper or computer and they should record any electrical maintenance and testing that has been carried out. The results of any tests should be recorded and the results should be made available for scrutiny.

Unless the circumstances make it unavoidable (for example, if an installer has ceased trading prior to certifying an installation), a Periodic Inspection Report should not be issued by one contractor as a substitute for an Electrical installation Certificate for work carried out by another contractor.

A Periodic Inspection Report does not provide a declaration by the designer or installer that the aspects of the work for which they were responsible comply with BS 7671. Also cables that are designed to be concealed cannot be inspected when construction is complete.

Completion Certificates and Inspection Report Forms

A Completion Certificate and an Inspection Report Form must be provided by the person responsible for the construction of the installation, or alteration thereto, or by an authorised person acting for them. Details of these certificates are given in IEE Regulations Appendix 6. The person who carries out any installation work assumes a very great responsibility in ensuring that the certificates are completed and that their terms are complied with in every respect. Any loss or damage incurred due to any neglect on the part of the person responsible for the installation might well involve claims for heavy damages.

Notice of Re-inspection and Testing

IEE Regulation 514.12.1 states that a notice, of such durable material as to be likely to remain easily legible throughout the life of the installation, shall be fixed in a prominent position at or near the main distribution board on completion of the work. It shall be inscribed as detailed within the regulation, in characters not smaller than those illustrated in Fig. 16.12.

IMPORTANT

This installation should be periodically inspected and tested and a report on its condition obtained, as prescribed in BS 7671 Requirements for Electrical Installations published by the Institution of Electrical Engineers.

Date of last inspection ....

Recommended date of next inspection ....

FIGURE 16.12 Wording specified by the IEE Regulations for the periodic inspection notice.

The determination of the frequency of periodic inspection is covered by IEE Regulation 622. No specific period is laid down, and an assessment needs to be made as to the use of the installation, the likely frequency of maintenance, and the possible external influences likely to be encountered. The person carrying out the inspection and testing, and completing the inspection certificate needs to take account of these issues. In the absence of other local or national regulations, a maximum period of five years would be applied, with shorter periods where appropriate.

All inspection and testing and the final results are required to be signed by a qualifying supervisor.

Roles of a Qualifying Supervisor

The qualifying supervisor must ensure, without any doubt, that before they sign an electrical installation certificate that the electrical installation complies with the building regulations, British Standards and the IEE Regulations.

The qualifying supervisor, so as to ensure without any doubt that the electrical installation being undertaken complies with the appropriate legislation, is advised to visit the site where the electrical installation is being undertaken.

The qualifying supervisor is advised to inspect the electrical installation at various stages, including:

a. During the installation of containment and cabling,
b. During the period of dead testing ($R_1 + R_2$ etc.),
c. During second fix of electrical items (socket outlets and accessories etc.) and
d. At the completion of the electrical installation so as to verify final live testing.

The qualifying supervisor at each stage is advised to communicate with the electrical designer so as to compare and discuss the results obtained and/or problems that may have occurred during the erection of the electrical services in question.

By using this method it may well quicken the process of finalising the project and help to solve any installation queries, ensuring that the electrical installation complies with current legislation before completion of the project. It is essential that the qualifying supervisor communicates with the contracts engineer and design engineer responsible for the project before, during and on completion of the project.

Once the qualifying engineer is satisfied that the installation complies with current legislation, the electrical certificate may be signed. A copy of the certificate must then be kept in a safe place with relevant information on the project undertaken including the periodic report notes of the qualifying supervisor during subsequent visits. The certificates must be made available throughout the year for perusal and inspection upon visits from the relevant inspection body, i.e. NICEIC and/or ECA.

Certificates

All certificates must be logged into the system before they can be distributed to the project engineers, each engineer is required to sign for the particular electrical installation certificate or book.

Appendix A – Extracts from IEE Tables

EXTRACT FROM IEE TABLE 41.3 – MAXIMUM EARTH FAULT LOOP IMPEDANCE (Z_S) FOR CIRCUIT-BREAKERS

Maximum Earth Fault Loop Impedance (Z_s) for Circuit-Breakers with U_o of 230V, for Instantaneous Operation Giving Compliance with the 0.4s Disconnection Time of Regulation 411.3.2.2 and 5s Disconnection Time of Regulation 411.3.2.3

(a) Type B circuit-breakers to BS EN 60898 and the overcurrent characteristics of RCBOs to BS EN 61009-1

Rating (A)	3	6	10	16	20	25	32	40
Z_s (ohms)	15.33	7.67	4.60	2.87	2.30	1.84	1.44	1.15

(b) Type C circuit-breakers to BS EN 60898 and the overcurrent characteristics of RCBOs to BS EN 61009-1

Rating (A)	6	10	16	20	25	32	40
Z_s (ohms)	3.83	2.30	1.44	1.15	0.92	0.72	0.57

Note: The circuit loop impedances given in the table should not be exceeded when the conductors are at their normal operating temperature. If the conductors are at a different temperature when tested, the reading should be adjusted accordingly.

CABLE COLOURS, INCLUDING EXTRACTS FROM IEE TABLE 51

Function	Alphanumeric	Colour (IEE Table 51)	Old fixed wiring colour (see text)
Protective conductors		Green and yellow	Green and yellow
Functional earthing conductor		Cream	Cream
a.c. Power circuit (including lighting)			
Phase of single-phase circuit	L	Brown	Red
Phase 1 of three-phase circuit	L1	Brown	Red
Phase 2 of three-phase circuit	L2	Black	Yellow
Phase 3 of three-phase circuit	L3	Grey	Blue
Neutral for single- or three-phase circuit	N	Blue	Black
Two-wire unearthed d.c. circuit			
Positive	L1	Brown	Red
Negative	L2	Grey	Black
Two-wire earthed d.c. circuit			
Positive (of negative earthed) circuit	L1	Brown	Red
Negative (of negative earthed) circuit	M	Blue	Black
Positive (of positive earthed) circuit	M	Blue	Black
Negative (of positive earthed) circuit	L2	Grey	Blue
Three-wire d.c. circuit			
Outer positive of two-wire circuit derived from three-wire system	L1	Brown	Red
Outer negative of two-wire circuit derived from three-wire system	L2	Grey	Red
Positive of three-wire circuit	L1	Brown	Red
Mid wire of three-wire circuit	M	Blue	Black
Negative of three-wire circuit	L2	Grey	Blue
Control circuits, extra-low voltage etc.			
Phase conductor	L	Brown, black, red, orange, yellow, violet, grey, white, pink or turquoise	
Neutral or mid wire	N or M	Blue	

EXTRACT FROM IEE TABLE 53.2 – SWITCHING DEVICES

An Extract from IEE Table 53.2 Showing Some Switching and Other Devices Permissible or the Purposes Shown. IEE Regulations Section 537 Gives Additional Information on this Topic

Device	Use as isolation	Emergency switching	Functional switching
RCD	Yes[a]	Yes	Yes
Isolating switch	Yes	Yes	Yes
Semiconductors	No	No	Yes
Plug and socket	Yes	No	Yes[b]
Fuse link	Yes	No	No
Circuit breaker	Yes[a]	Yes	Yes
Cooker control switch	Yes	Yes	Yes

[a]Provided device is suitable and marked with symbol per BS EN 60617.
[b]Only if for 32A or less.

EXTRACT FROM IEE TABLE 4B1 – TEMPERATURE RATING FACTORS

Rating Factors for Ambient Air Temperatures Other Than 30 °C to be Applied to the Current-Carrying Capacities for Cables in Free Air

| Ambient temperature (°C) | Insulation | |
	70 °C Thermoplastic	90 °C Thermosetting
25	1.03	1.02
30	1.00	1.00
35	0.94	0.96
40	0.87	0.91
45	0.79	0.87
50	0.71	0.82
55	0.61	0.76
60	0.50	0.71

EXTRACT FROM IEE TABLE 4C1 – GROUPING RATING FACTORS

Rating Factors for One Circuit or One Multicore Cable or for a Group of Circuits, or a Group of Multicore Cables, to Be Used with Current-Carrying Capacities of Tables 4D1A–4J4A

	Number of circuits or multicore cables							
Arrangement (cables touching)	1	2	3	4	5	6	7	8
Bunched in air, or on a surface, embedded or enclosed, methods A–F	1.00	0.80	0.70	0.65	0.60	0.57	0.54	0.52
Single layer on wall or floor, method C	1.00	0.85	0.79	0.75	0.73	0.72	0.72	0.71

Extracts from notes:
These factors are applicable to uniform groups of cables, equally loaded.
Where horizontal clearances between adjacent cables exceed twice their overall diameter, no rating factor need be applied.
When cables having different conductor operating temperatures are grouped together, the current rating is to be based upon the lowest operating temperature of any cable in the group.
If, due to known operating conditions, a cable is expected to carry not more than 30% of its *grouped* rating, it may be ignored for the purpose of obtaining the rating factor for the rest of the group.

EXTRACT FROM IEE TABLE 4D1A – CURRENT-CARRYING CAPACITY, COPPER CONDUCTORS

Single-Core 70 °C Thermoplastic Insulated Cables, Non-Armoured, with or without Sheath Ambient Temperature 30 °C, Conductor Operating Temperature 70 °C

	Reference method A (enclosed in conduit in thermally insulating wall etc.)		Reference method B (enclosed in conduit on a wall or in trunking etc.)		Reference method C (clipped direct)	
Conductor cross-sectional area	Two cables, single-phase a.c. or d.c.	Three or four cables, three-phase a.c.	Two cables, single-phase a.c. or d.c.	Three or four cables, three-phase a.c.	Two cables, single-phase a.c. or d.c. flat and touching	Three or four cables, three-phase a.c. flat and touching or trefoil
1	2	3	4	5	6	7
mm^2	A	A	A	A	A	A
1	11	10.5	13.5	12	15.5	14
1.5	14.5	13.5	17.5	15.5	20	18
2.5	20	18	24	21	27	25
4	26	24	32	28	37	33
6	34	31	41	36	47	43
10	46	42	57	50	65	59
16	61	56	76	68	87	79
25	80	73	101	89	114	104
35	99	89	125	110	141	129

EXTRACT FROM IEE TABLE 4D1B – VOLTAGE DROP (PER A PER M), COPPER CONDUCTORS

Conductor Operating Temperature 70 °C

		Two cables, single-phase a.c.		
			Reference methods C and F (clipped direct, on tray or in free air)	
Conductor cross-sectional area	Two cables d.c.	Reference methods A and B (enclosed in conduit or trunking)	Cables touching	Cables spaced[a]
1	2	3	4	5
mm^2	mV/A/m	mV/A/m	mV/A/m	mV/A/m
1	44	44	44	44
1.5	29	29	29	29
0.5 2.5	18	18	18	18
4	11	11	11	11
6	7.3	7.3	7.3	7.3
10	4.4	4.4	4.4	4.4
16	2.8	2.8	2.8	2.8
		$r \times z$	$r \times z$	$r \times z$
25	1.75	1.80 0.33 1.80	1.75 0.20 1.75	1.75 0.29 1.75
35	1.25	1.30 0.31 1.30	1.25 0.28 1.25	1.25 0.28 1.30
50	0.93	0.95 0.30 1.00	0.93 0.19 0.95	0.93 0.28 0.97

[a]Spacings larger than one cable diameter will result in a larger voltage drop.

EXTRACT FROM IEE TABLE 4E4A – CURRENT-CARRYING CAPACITY, COPPER CONDUCTORS

Multicore 90 °C Armoured Thermoplastic Insulated Cables

Conductor cross-sectional area	Reference method C (clipped direct)		Reference method E (in free air or on a perforated cable tray etc., horizontal or vertical)		Reference method D (direct in ground or in ducting in ground, in or around buildings)	
	One two-core cable, single-phase a.c. or d.c.	One three- or four-core cable, three-phase a.c.	One two-core cable, single-phase a.c. or d.c.	One three- or four-core cable, three-phase a.c.	One two-core cable, single-phase a.c. or d.c.	One three- or four-core cable, three-phase a.c.
1	2	3	4	5	6	7
mm^2	A	A	A	A	A	A
1.5	27	23	29	25	25	21
2.5	36	31	39	33	33	28
4	49	42	52	44	43	36
6	62	53	66	56	53	44
10	85	73	90	78	71	58
16	110	94	115	99	91	75

Notes:
Where a conductor operates at a temperature exceeding 70 °C it must be ascertained that the equipment connected to the conductor is suitable for the conductor operating temperature (see Regulation 512.1.2).
Where cables in this table are connected to equipment or accessories designed to operate at a temperature not exceeding 70 °C, the current ratings given in the equivalent table for 70 °C thermoplastic insulated cables (Table 4D4A) must be used (see also Regulation 523.1).

EXTRACT FROM IEE TABLE 4E4B – VOLTAGE DROP (PER A PER M), COPPER CONDUCTORS

Conductor Operating Temperature 90 °C

Conductor cross-sectional area 1	Two-core cable, d.c. 2	Two-core cable, single-phase a.c. 3			Three- or four-core cable, three-phase a.c. 4		
mm²	mV/A/m	mV/A/m			mV/A/m		
1.5	31	31			27		
2.5	19	19			16		
4	12	12			10		
6	7.9	7.9			6.8		
10	4.7	4.7			4.0		
16	2.9	2.9			2.5		
		r	x	z	r	x	z
25	1.85	1.85	0.160	1.90	1.60	0.140	1.65
35	1.35	1.35	0.155	1.35	1.15	0.135	1.15
50	0.98	0.99	0.155	1.00	0.68	0.135	0.87

Glossary of Terms

A:	Amperes (current)
ACB:	Air circuit breaker
ADMD:	After diversity maximum demand
AL:	Aluminium conductor
Applicant:	The company wishing to undertake the contestable work
AWA:	Aluminium wire armoured
BNO:	Building Network Operator - The organisation that owns or operates, by permission of licence or licence exemption, the electricity distribution network within a multiple occupancy building, between the intake position and customers' installations
BNO main:	BNO cable (or busbar) which connects more than one customer
BNO network:	Cables (or bus-bars) switch/fusegear and ancillaries between the intake position and customer's premises
BNO service:	BNO cable which connects a single customer
BS:	British Standard
BS EN:	A European Standard adopted as a British Standard
BSI:	British Standards Institution
CB:	Circuit breaker
CNE:	Combined neutral and earth (of cable construction)
CPC:	Circuit protective conductor
CPE cable:	Chlorinated polyethylene
CSP cable:	Chlorosulphonated polyethylene
Customer:	The owner or tenant of a defined property, each having its own metering point, housed within a larger building
Customer's installation:	The electrical system installed within and servicing an individual customer's premises
DB:	Distribution Board
DLH:	Distribution Licence Holder – defined in Standard Licence Conditions for Electricity Distributors, issued under the Utilities Act and effective from 1st Sept. 2001
DNO:	Distribution Network Operator - The organisation that is licensed or permitted by licence exemption to operate a public Distribution Network and is responsible for confirming requirements for the connection of equipment to that network
DSA:	Distribution Service Area – the service area of a DLH
EA:	Electricity Association (replaced by ENA for Networks issues post Oct 2003)
EMC:	Electromagnetic compatibility
ENA:	Energy Networks Association
ENATS:	Energy Networks Association Technical Specification
EPR cable:	Ethylene propylene rubber
ESQCR:	The Electricity Safety Quality & Continuity Regulations 2002
ETFE cable:	Ethylene tetrafluoroethylene

FP cable:	Fire performance
FR cable:	Fire retarded
HD:	Harmonised Document (IEC standard adopted as a European reference document)
HOFR cable:	Heat oil resistant and flame retardant
Host DLH:	The DLH in whose licensed area (DSA) the works are to take place
HSE:	Health & Safety Executive
HV:	High voltage; ie a voltage exceeding 1000V a.c.
IEC:	International Electrotechnical Commission
IET:	Institution of Engineering and Technology
Intake position:	The location within the building where the boundary between the Distribution Network Operator's network and the Building Network Operator's network occurs
Intake terminals:	The electrical point of connection between the Distribution Network Operator's network and the Building Network Operator's network
kW:	Kilowatt
Landlord:	The owner of a multiple occupancy building, who may be other than the BNO
LDPE cable:	Low density polyethylene
LSF cable:	Low smoke and fumes Made to BS 6724
LSOH cable:	Low smoke, zero halogen
LV:	Low voltage, ie a voltage not exceeding 1000V a.c.
m:	Metre
MCB:	Miniature circuit breaker
MCCB:	Moulded case circuit breaker
Metering point:	A point at which settlement metering is installed
MI cable:	Mineral insulated
MOCOPA:	Meter Operators Code of Practice Agreement
MOP:	Meter operator - An appointed agent for installing and maintaining electricity metering equipment
MPAN:	Metering Point Administration Number - A unique number provided for each metering point by the relevant network operator
Multiple occupancy building:	A building occupied by more than one customer
NH cable:	Non-halogenated Made to BS 6724
NRSWA:	New Roads and Street Works Act
OFGEM:	Office of Gas and Electricity Markets
PELV:	Protective extra-low voltage
PILC cable:	Paper insulated lead covered Sometimes with PVC sheath
PME:	Protective multiple earthing
PSCC:	Prospective short circuit current
PTFE cable:	Polytetrafluoroethylene
PVC cable:	Polyvinylchloride Made to BS 6346
PVC/SWA/PVC:	Cable with PVC inner and outer insulation and steel wire armouring
RS cable:	Reduced smoke
SELV:	Separated extra-low voltage
Supplier:	An organisation which contracts with customers to supply electrical energy
Supply terminals:	With respect to a multiple occupancy building the supply terminals shall be the final terminals of the metering systems of each metering point connecting to customers' installations
SWA cable:	Steel wire armoured
Tri-rated cable:	PVC insulated cables for switchgear and control wiring complying with three standards: (1) Type CK cables to BS 6231 (2) Type TEW equipment wires to Canadian Standard C22.2 No 127. (3) American Underwriters Laboratories (UL) Subject 758.
TRS cable:	Tough rubber sheath
V:	Voltage or Volts
VCB:	Vacuum circuit breaker
VR cable:	Vulcanised rubber
XLPE cable:	Crosslinked polyethylene suitable for 90 °C. Made to BS 5467
Ze:	Earth fault loop impedance